A MOST ELEGANT EQUATION

ALSO BY DAVID STIPP

The Youth Pill: Scientists at the Brink of an Anti-Aging Revolution

A Most Elegant Equation

EULER'S FORMULA AND THE BEAUTY OF MATHEMATICS

DAVID STIPP

BASIC BOOKS
New York

Basic Books
Hachette Book Group
1290 Avenue of the Americas, New York, NY 10104
www.basicbooks.com

Printed in the United States of America

First Edition: November 2017

Published by Basic Books, an imprint of Perseus Books, LLC,
a subsidiary of Hachette Book Group, Inc.

Library of Congress Cataloging-in-Publication Data has been applied for.
ISBNs: 978-0-465-09377-9 (hardcover), 978-0-465-09378-6 (e-book)

LSC-C

10 9 8 7 6 5 4 3 2 1

To Alicia, Quentin, and Claire

Can you recall why you fell in love with mathematics? It was not, I think, because of its usefulness in controlling inventories. Was it not instead because of the delight, the feelings of power and satisfaction it gave; the theorems that inspired awe, or jubilation, or amazement; the wonder and glory of what I think is the human race's supreme intellectual achievement?

UNDERWOOD DUDLEY,
DePauw University Emeritus
Professor of Mathematics

CONTENTS

A MOST ELEGANT EQUATION

Introduction

Who can resist lists of the ten best this or that? Not me. So a few years ago when I ran across a ranking by math experts of the most beautiful theorems, I tuned in. I decided to treat it as a pop quiz: How many of them could I dredge up from my distant undergraduate days as a math major? Alas, I'm afraid I flunked. Still, I was consoled by being able to recall nine of the top 10 (of two dozen). But the No. 1 beauty, an equation known as Euler's formula, bothered me—I'd seen it before, yet couldn't remember going over it in school.

It's probably overrated, I thought, a little defensively. Lacking arcane symbols or other bona fides of serious mathematical artistry, it features only numbers, and just five at that (as it's usually written). True, three of them are designated by letters, showing they're special. But the equation* itself looks hardly more scintillating than a confused first grader's $2 + 1 = 0$. See: $e^{i\pi} + 1 = 0$.

I was certainly familiar with Leonhard Euler (pronounced "oiler"), the eighteenth-century mathematician it's named after. He's known as the Mozart of mathematics, and his fingerprints,

*Euler's equation is often called a formula or an identity. I refer to it as an equation or a formula, glossing over the fact that these terms aren't synonymous in formal mathematics.

so to speak, were all over the pages of my old math books. But that didn't tell me much. And as the formula started playing through my head like a tune whose provenance maddeningly eluded me, I discovered via Google that a number of authorities on math have considered it not only beautiful, but also one of the most remarkable results in the history of mathematics. Among them was one of my heroes, Richard Feynman, a brilliant theoretical physicist who worked on the Manhattan Project, won a Nobel Prize, led the investigation of the 1986 Space Shuttle Challenger disaster, and, to top it off, radiated almost superhuman *joie de vivre*. Why did this simple-looking little formula light up his multi-gigawatt mind?

OK, I decided, it's time for some investigative reporting. Although I'd moved on to science writing after college, I'd warded off total math-muscle atrophy by acting as my eye-rolling kids' tragicomically enthused mathematics tutor, all the way through high school calculus in the case of my son. So I looked up the derivation of Euler's formula—it's straightforward if you know a little calculus—and reconnoitered its history and significance. And like many math lovers before me, I came away thinking, "wow," or, more precisely, "WOW!"

For one thing, it effectively compresses about two millennia's worth of big ideas in mathematics into a fantastically small package, among them the nature and uses of infinity (∞ is basically tucked away inside the formula), the weird ubiquity of the number π in math, the great utility of the misleadingly named imaginary numbers, and the wonderfulness of nothing, i.e., zero. What really grabbed me, though, was the fact that on his way to the formula, Euler uncovered a set of hidden connections among math concepts that many students go over in high school without ever realizing that they're deeply linked in a way that could aptly be described as scary-cool. (That is, several levels of cool up from merely awesome.)

So I got the beauty thing. But I still wondered about the blank space in my memory where I should have had a beauty queen. Continuing the investigation, I pulled out my college calculus books, which I'd kept as trophies for all the hours I spent hunched over them. Euler's formula wasn't listed in their indexes. Paging through them, I finally found a single, fleeting mention of a general equation (also Euler's) from which the most beautiful formula is derived as a special case. The closest thing to "$e^{i\pi} + 1 = 0$" that I could find was an exercise whose answer happened to be a version of it.

So that's it, I thought. I didn't forget Euler's formula. It's just that, inexplicably, it got infinitesimal shrift during my student days. Come to think of it, the most beautiful equation didn't come up in any of my son's high school math courses either.

This last thought led to my recollecting how I knew all too much about those courses. It's true, I inwardly sighed, as Quentin's tutor I was more on top of his daily math assignments than he was. A budding artist, he regarded math classes as a boring waste of time. And by the time he left for college, I was seeing double when I looked at his math books. One of the two images I saw was the engaging one I knew as a math lover. But the other image was the one that I knew he saw. Novelist Nicholson Baker memorably described it in a 2013 piece in *Harper's* about how mandatory high-school math classes often foster math hate. In particular, he wrote, Algebra II students "are forced, repeatedly, to stare at hairy, square-rooted, polynomialed horseradish clumps of mute symbology that irritate them, that stop them in their tracks, that they can't understand. The homework is unrelenting, the algorithms get longer and trickier, the quizzes keep coming. Sooner or later, many of them hit the wall."

You can probably see where I'm going with this. My next thought was: If only Quentin, as well as millions of other people out there who regard math as the supreme soporific, could experience the frisson I felt when reliving Euler's great discovery. But let's get real, I firmly told my inner soliloquist, who was already whispering, "Do a book!" in my ear. There's no way that people who have forgotten most of their high school math could experience that epiphany. Preposterous. Forget it. The very idea could do incalculable harm to my reputation as a somewhat grounded thinker.

And then, of course, I sat down to write this book. Well, that's a slight exaggeration. I mulled the idea over for about a year. Finally I remembered how American philosopher Oets Kolk "O.K." Bouwsma had overcome his own hesitation to put together a book he'd had doubts about: "I have tossed a coin and it came down as I thought it would," he explained. "It stood on its edge. And I knocked it down."

So here we are. But before you delve into this book—or turn away from it—please finish the intro. I'll be brief.

One reason I flattened the coin is that Euler's formula offers a very rare combination of beauty, depth, surprise, and, most importantly for my purposes, understandability—few profound math results are as accessible as it is. (Although it does take some explaining—a short book's worth, apparently.) And I knew that writing about it would give me license to roam across the history of mathematics as I went over the ideas that it so cunningly encapsulates.

Also, the formula isn't just a math version of abstract art. Long after Euler's era, scientists and engineers realized that the general equation mentioned above (the conceptual parent of $e^{i\pi} + 1 = 0$) is immensely useful for mathematically modeling phenomena, such as the rhythmic flow of alternating current. Thus, Euler's brilliant pure-math discovery is now effectively embedded in electrical devices all around us. I'd call that scary-cool, too, if I hadn't already used up my weekly allotment of that epithet.

The formula's timelessness also appealed to me. As electrical engineer Paul Nahin nicely put it in a book he wrote about the equation for people who've taken college math: "Unlike the physics or chemistry or engineering of today, which will almost surely appear archaic to technicians of the far future, Euler's formula will still appear, to the arbitrarily advanced mathematicians ten thousand years hence, to be beautiful and stunning and untarnished by time."

I hope this book is approximately equal to the sum of these pluses. But I feel obliged to mention a couple of things up front by way of truth in advertising. First, it isn't intended to increase quantitative skills or impart a thorough grounding in the math that it covers—its sole mission is to bring home that great mathematics is as provocative, beautiful, and deep as great art or literature. Second, if you're a long-time math lover, you'll probably find most of it too elementary (with the possible exception of Appendix 1). I've tried to make the book's math accessible to those who have forgotten, or perhaps repressed, most of the mathematics they learned after sixth grade. Thus, I assume that readers are acquainted with little more than the basics needed to cope in life: arithmetic, fractions, ratios, decimals, percentages—essentially what American kids are expected to know before being introduced to algebra in seventh or eighth grade.

Still, I have included a number of equations (along with lots of hand-holding as they rear their Medusa-like heads). You might conclude from this that I somehow missed physicist Steven Hawking's famous admonition about the use of equations in nontechnical books: "Someone told me," he remarked, "that each equation I included in the book [*A Brief History of Time*] would halve the sales. I therefore resolved not to have any equations at all." (In the end he did include one: $E = mc^2$.) Actually, I'm so acutely aware of this comment that it's virtually tattooed on my cerebral cortex. But I decided that giving an account of Euler's formula without equations would be like describing van Gogh's *The Starry Night* without showing a good-sized image of it. It would defeat my purpose, which is to enable readers to make an authentic personal approach to a high point in the history of mathematics, and, indeed, of human thought.

It has occurred to me, of course, that if the halving principle is correct this book is likely to attract less than one-millionth of a reader. (There are several dozen equations.) I'm happy to report, however, that more than a million times that many people are now known to have dipped into it (meaning you, dear whole number of a reader), which suggests that the principle's originator may need remedial math. In any case, I hope that at least a few hungry-minded persons will read on and find themselves experiencing some unexpected moments of amazement and jubilation. What better reason to read a book? Or, for that matter, to write one?

CHAPTER 1

God's Equation

Leonhard Euler seemed as curious, cheerful, and acute as ever during the morning of September 18, 1783, the last day of his life. It was nearly fall in St. Petersburg, Russia, where the 76-year-old Swiss mathematician lived with his extended family and worked at the Russian Academy of Sciences. Despite having been nearly blind for more than a decade, he'd continued to produce math and science papers at an astounding rate—he'd actually become more productive after losing his sight. Euler did complex calculations in his head and dictated the results to assistants, who recorded them on two large writing slates he kept in his study.

That morning, as was his custom, he gave one of his grandsons a lesson in elementary science. Soon after, two colleagues arrived to discuss scientific matters, including the recently discovered planet Uranus and innovative hot-air balloons that had been launched in unpiloted experiments a few months earlier by two French brothers, Joseph-Michel and Jacques-Étienne

Montgolfier—they would soon go down in history for achieving the first manned ascent.

After informing his visitors that he'd lost his remaining vestige of vision, Euler proceeded to work out a complex differential equation (an exercise in calculus) to model the rise of hot-air balloons and thereby determine how high they could go. He also made mental calculations related to Uranus's orbit. After lunch, he said he felt faint and lay down for a nap.

A few hours later he rejoined his family and friends for four o'clock tea. Sitting on a couch, he jovially played with one of his grandsons, and then asked his wife for a second cup. Two minutes later, he suddenly dropped the pipe he was smoking, stood up, clasped both hands to his forehead, and exclaimed, in German, "I am dying." They were his last words, and they represented one of his characteristic flashes of deep intuition—he was suffering a stroke that would prove fatal. Soon losing consciousness, he died that evening.

Later that day, Johann, Euler's eldest son, came across the latest calculations on his father's slates. Despite his shock and sadness (or perhaps because of them) he quickly set to work fleshing out the ones on hot-air balloons and published the results in a French journal as Euler's first posthumous paper. Many followed—Euler left a huge collection of unpublished manuscripts filled with important findings that came out for decades after his death.

EULER'S LIFE SPANNED most of the Enlightenment, an explosive efflorescence of thought that followed the spring-like flowering of the Renaissance. During its high time in the eighteenth century, intellectuals gathered in coffeehouses and literary salons to thrash out world-changing ideas about science, individual liberty, religious tolerance, and free-market economics. America's soul was forged during the Enlightenment—Jefferson was channeling its spirit when he penned the Declaration of Independence in 1776. That same year, Scottish philosopher Adam Smith published *The Wealth of Nations*, which spread the enormously influential idea that rational self-interest can foster economic prosperity. A few years later, England's Mary Wollstonecraft wrote *A Vindication of the Rights of Woman*, one of the earliest feminist tracts.

The Enlightenment's great minds included George Frideric Handel, Wolfgang Amadeus Mozart, Franz Joseph Haydn, Jonathan Swift, Alexander Pope, Samuel Johnson, Daniel Defoe, Voltaire, Thomas Paine, Benjamin Franklin, Montesquieu, David Hume, Immanuel Kant, Denis Diderot, William Herschel, Antoine Lavoisier, and Émilie du Châtelet, a mathematician and physicist who was the first woman to have a scientific paper published by the French Academy of Sciences. Many were part of an international republic of letters whose unofficial motto, famously enunciated by Kant, was *Sapere aude*! (Latin for "Dare to know!"). The era's most influential writers, such as Swift, Paine, and Voltaire, attacked received wisdoms and the powers-that-be with unprecedented gusto and wit. A number of them regarded human reason, not divine revelation, as the ultimate arbiter of truth. That led many of the intellectuals of the time to embrace deism, which held that God set up the universe to run according to discoverable natural laws and then permanently retired from the scene;

deists rejected miracles and other supernatural phenomena as superstitions.

Some, following England's John Locke, believed that natural law applied to human society, and among its dictates were the rights to life, liberty, and property—a truly revolutionary idea. It was also the age of "enlightened despots," such as Prussia's King Frederick II and Russia's Empress Catherine II, who shared the era's belief in the primacy of reason, and who fostered education, religious tolerance, and property rights. Frederick and Catherine also took pride in their countries' scientific academies, which vied to attract Europe's most brilliant minds and served as leading research centers. Euler began his career at the Russian academy and then spent a quarter century at the Prussian academy before returning to the one in Russia for the rest of his life.

The leading thinkers of the time often disagreed, and the big questions they took up are still being debated. Euler, for example, was more religiously conservative than many of his peers, and he rejected the deists' then-radical elevation of human reason over Christian revelation. But almost all of the era's greats, including Euler, shared the Enlightenment's defining inspiration: that the world is governed by natural laws whose regularities can be discovered by the methods of science. This idea was closely allied with the view that nature's laws can be expressed in mathematical terms, a concept that seemed to be written in the very stars after Isaac Newton showed during the late 1600s that everything from the tides to the planets' trajectories across the sky could be explained in terms of mathematically-formulated laws of motion and gravitation. The energetic pursuit of math-based explanations during the Enlightenment paved the way for the Industrial Revolution and the rise of modern technology. This pursuit also informed the period's formal style of discourse, which sometimes had a distinctly mathematical flavor. Consider the Declaration of Independence's famous second sentence: "We hold these truths to be self-evident, that all men are created equal, that they are endowed by their Creator with certain unalienable Rights, that among these are Life, Liberty and the pursuit of Happiness"—it has the ring of a mathematical axiom from which the document's subsequent assertions logically follow like theorems.

Such ornate, well-knit prose may seem quaint in the age of tweets and blogs. But it is difficult to overstate the importance of the passion for balance, order, and reason that informed Jefferson's words and that Euler's era bequeathed to posterity. Indeed, when we glance back through the corridors of history, the Enlightenment stands out like a brightly illuminated room whose glow is still helping us make our way forward in the dark. And as I write this in late 2016, it seems that today's post-truth autocrats, religious fundamentalists, and science deniers have joined forces to pursue a single terrible end: to extinguish its guiding light and replace it with greed, lies, ignorance, and hate.

EULER WAS THE ENLIGHTENMENT'S greatest mathematician, as well as the era's greatest physicist. He also made many important contributions to astronomy and engineering. Science historian Clifford Truesdell has estimated that Euler wrote a *quarter* of all the mathematical and scientific works published during the eighteenth century. During most of the Enlightenment he was, in effect, the leading carrier of the torch that had been lit during the Renaissance, and that Newton and others had raised high during the late seventeenth century.

Euler won the French science academy's annual prize for innovative solutions to science, mathematics, and technology problems—the eighteenth century's version of today's Nobel Prize—no less than 12 times. He was history's most prolific mathematical innovator; his collected works, which fill 80 thick volumes (so far), are math's Mt. Everest. And he made major advances across virtually every subdiscipline of mathematics. As mathematician William Dunham has noted, you'd need a forklift to transport the roughly 25,000 pages of his *Opera Omnia* (Euler's complete works, which the Swiss Academy of Sciences has been compiling for over a century).

Euler also made technology advances. In 1752, for instance, he originated then-futuristic designs for a paddle wheel and a rotating propeller for ships—devices that wouldn't be practical until steam engines were developed in the nineteenth century to turn them. His ideas on how to construct achromatic lenses helped inspire an English inventor to put together one of the first such color-correcting lenses. He even proposed a design for a logic machine—a kind of early, mechanical computer—although it's not clear whether the machine was constructed.

Although Euler isn't considered a philosopher, his writings clearly influenced Kant's metaphysics. His work on the mathematics of population growth helped inspire Darwin to come up with the theory of natural selection. He had a hand in solving the famous longitude problem—the devising of a practical way for determining a ship's longitude at sea—and in 1765 the British government awarded him part of the prize money it had offered in 1714 to spur inventors to attack the problem. Today, the mathematics he pioneered has been applied in systems used for secure Internet communications, the analysis of social networks, electronic circuit design, and many other things. Even Hollywood has recognized his contributions—the 2016 movie *Hidden Figures* mentioned "Euler's method," an approximating technique one of the film's stars used to calculate trajectories of 1960s spacecraft so they didn't burn up when re-entering the atmosphere and then splashed down where Navy ships could quickly pick them up.

It's clear that Euler had one of the most beautiful minds in history. And unlike many great innovators, he achieved international fame during his lifetime—perhaps only Voltaire, with his ebullient irreverence and memorably caustic wit, was more renowned in Europe than he was during the late 1700s. Today, however, while many people are familiar with the works of other Enlightenment geniuses—Handel's music, Voltaire's satires, Defoe's novels, to name a few examples—Euler's works are known to relatively few. That's a shame. The intellectual gems he carved are as scintillating as any to be found in the history of thought. One, in particular, stands out as the epitome of math's unsurpassed power to surprise and delight.

IT IS AS GALVANIZING, puzzling, and concise as a Zen koan—math's one hand clapping:

$$e^{i\pi} + 1 = 0.$$

Mathematics textbooks call it Euler's formula. But some people feel that that name is too mundane for what is arguably math's most magnetic truth, as well as one of its most startling, and so they call it God's equation.*

Benjamin Peirce, considered America's first world-class mathematician, once showed how to prove a variation of it during a class he taught at Harvard, where he was a professor from 1831 to 1880. Then, after contemplating what he'd written on the chalkboard for a few minutes, he turned to his students and observed that it "is absolutely paradoxical; we cannot understand it, and we don't know what it means, but we have proved it, and therefore we know it must be the truth."

Mathematician Keith Devlin, "The Math Guy" on National Public Radio, has been bowled over too. "Like a Shakespearean sonnet that captures the very essence of love, or a painting that brings out the beauty of the human form that is far more than just skin deep, Euler's equation reaches down into the very depths of existence," he commented.

Richard Feynman was briefer but no less enthusiastic: It's "the most remarkable formula in math," he stated in a notebook he compiled at age 14, along with a sketch of its proof. As mentioned in the introduction, math experts have ranked Euler's formula as mathematics' most beautiful equation. And in 2014 fifteen mathematicians were presented with 60 different equations, including Euler's formula, while undergoing brain scans; the study showed that Euler's formula had the

*The nickname "God's equation" is meant to underscore its profundity, not to suggest that a deity handed it down from on high. However, it is true that mathematician Jules Henri Poincaré once called Euler the "god of mathematics." And if God existed, my guess is that she would have Euler's formula, carved on a stone tablet, sitting on her desk.

greatest ability to activate brain areas associated with experiencing visual and musical beauty.

On its far left side are numbers of infinite complexity (two of them, e and π, are designated with letters because it would literally take forever to write them down with numerals), yet their endlessness collapses into a small, neat integer when they're combined. That is, the equation shows that $e^{i\pi}$—which seems the kind of numerical monstrosity that would turn a young math student to stone—is actually equal to a simple integer, −1.* Thanks to this startling fact, the equation's five seemingly unrelated numbers (e, i, π, 1, and 0) fit neatly together in the formula like contiguous puzzle pieces. One might think that a cosmic carpenter had jig-sawed them one day and mischievously left them conjoined on Euler's desk as a tantalizing hint of the unfathomable connectedness of things.

*To see this, it helps to remember that adding the same number to both sides of an equation (by "sides" I mean the stuff to the left and right of the equals sign) turns the two sides into other stuff that's equal. If −1 is added to both sides of Euler's formula, the left side becomes $e^{i\pi} + 1 + (-1)$, which equals $e^{i\pi}$ since $1 + (-1)$ equals 0, and adding 0 to a number leaves it just as it is. Meanwhile, the right side becomes $0 + (-1)$, which, since added 0s can be ignored, equals −1. Thus, this maneuver turns Euler's formula into $e^{i\pi} = -1$, which means that $e^{i\pi}$ is really just −1 in a fantastically good disguise.

CHAPTER 2

A Constant That's All About Change

At first glance, the number *e*, known in mathematics as Euler's number, doesn't seem like much. It's about 2.7, a quantity of such modest size that it invites contempt in our age of wretched excess and relentless hype. Puffy numbers like a bazillion clearly get a lot more ink in the media.* With *e* dollars in your pocket, you wouldn't even have enough to buy a small latte at my local Starbucks:

FIGURE 2.1

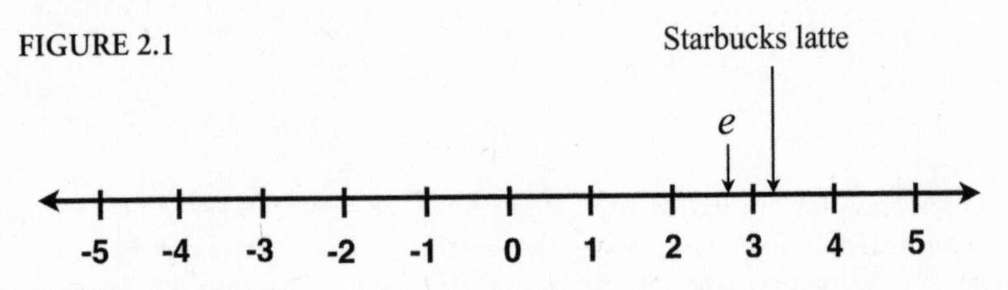

*A bazillion, like a zillion, is a fictitious large number of indeterminate size. So is a gazillion, which may or may not be a zillion times bigger than a bazillion.

But e is not to be trifled with. It's one of math's most versatile superheroes.

To begin with, it's uniquely valuable for mathematically representing growth or shrinkage. That alone makes it a standout. In fact, e's usefulness for dealing with problems related to the growth of savings via compound interest is what brought about its discovery in the 1600s.* Let's revisit that moment in history with a modern twist.

Say a newly formed bank decides to offer annual interest of 100% on savings accounts. (That's pretty farfetched, I realize, but it can happen in math fantasies.) Seeing its ads online, a neurotically cautious guy takes a single dollar from under his mattress and deposits it. A year later his savings would be equal to his original deposit, \$1, plus the interest. This year-end total can be calculated by multiplying his original deposit (the principal) times the quantity "$1 + r$," where r stands for the interest rate expressed as a decimal number. (The 1 in "$1 + r$" accounts for the principal that he'll get back at the end of the year, and the r accounts for the interest that he'll get in addition to the principal.) The total would therefore be $\$1 \times (1 + r)$. Since r equals 1.00 in this case (that is, 100% expressed as a decimal number), this would be $\$1 \times (1 + 1) = \$1 \times 2 = \$2$.

After a year it dawns on the guy that the bank isn't too big to fail, and so he hastily withdraws the \$2 and makes a beeline for his mattress. But the bank soon lures him back with an even better deal: Henceforth, the 100% annual interest will be broken into two 50% chunks paid after the first and second six months of the year.

---

*The number we now call e was known before Euler came along, but he brought it to the fore in mathematics and chose the symbol e for it.

Again, he deposits a single dollar. Six months later, his account would contain his original deposit plus the first 50% interest payment. Using the calculating method shown above, this would be $\$1 \times (1 + r)$, where r is one-half of 100% expressed as a decimal, or 0.50. If the updated amount is then treated as his new principal, his savings after the second interest payment at the end of the year would be equal to the new principal multiplied times the term $(1 + r)$ again. That is, it would be $[\$1 \times (1 + r)] \times (1 + r)$, which, with 0.5 subbed in for r, would be $\$1 \times 1.5 \times 1.5$, or \$2.25.

These calculations suggest a handy rule of thumb: The addition of compounding periods translates arithmetically into repeatedly multiplying by $1 + r$. Thus, if the bank divided its annual 100% interest into three equal interest payments compounded every third of a year, the guy's one-year total would be $\$1 \times (1 + r) \times (1 + r) \times (1 + r)$. The r in this case would be about 0.33, or one-third of 100% expressed as a decimal, and using that number for r yields a year-end total of about \$2.37.

If the bank got so desperate to attract depositors that it compounded quarterly, his one-year amount would be $\$1 \times (1 + r) \times (1 + r) \times (1 + r) \times (1 + r)$, with r equal to one-fourth of 100%, or, as a decimal, 0.25. This comes to about \$2.44.

There's a pattern developing here—it's actually a faint outline of *e*—but it's hard to see at this point. A little more compound-interest magic is needed to make it conspicuous.

But before proceeding, let's turn our rule of thumb into a simple formula in order to facilitate cranking out the guy's accumulated savings at the end of one year. To do that, we'll presume that the bank divides its 100% annual interest into n equal interest payments (where n stands for a whole number greater than zero) spread out through the year. That would mean that the interest rate for each compounding period would be 1.00 (again, that's 100% expressed as a decimal) divided by n, or $1/n$ when expressed as a fraction. Thus, we can replace the r in "$1 + r$" with $1/n$, turning the expression into "$1 + 1/n$." That means his one-year total can be written $\$1 \times (1 + 1/n)^n$, where the superscripted n is an **exponent.** The n tells how many times the term $(1 + 1/n)$ is multiplied times the original deposit of $1—using n as an exponent accounts for the fact that we must multiply once for each of the n compounding periods. And since multiplying any number times 1 leaves it unchanged, we can express the generalized year-end total, $\$1 \times (1 + 1/n)^n$, as simply $(1 + 1/n)^n$ dollars.

Now we're ready to make e jump out. We need only ask a simple question: If n is extremely large, will the guy realize his dream of hitting the jackpot by leaving his dollar in the bank for a year?

Exponents specify how many times to multiply a number by itself. For instance, 10^2, which is spoken "ten squared" or "ten raised to the second power," means 10×10, or 100. Likewise, 5^3, or "five raised to the third power," means $5 \times 5 \times 5$, or 125.

At first blush, you might think the answer is yes, because the above calculations showed that spreading the 100% annual interest over more compounding periods (equivalent to increasing n) led to increasing year-end totals. But there's a tight-fisted devil in the details here: The gains got progressively smaller as n got larger. For instance, as n was increased from one to two to three to four, the total went from $2 to

\$2.25 to \$2.37 to \$2.44. This shrinking-gains phenomenon continues with a vengeance: If the bank compounded weekly ($n = 52$) the total would be a little over \$2.69. If it compounded daily ($n = 365$), the guy would rack up about \$2.71. And if it compounded every second, he'd wind up with a smidgeon more than \$2.7182.

So much for his hitting the jackpot. Indeed, you might conclude, quite correctly, that as n becomes larger and larger, the expression for calculating his one-year total—$(1 + 1/n)^n$—will get ever closer to a number that's only a little greater than 2.718. Such boundary numbers are called limits in math, and in this case the limit is the number e. In fact, e is usually defined as the number that $(1 + 1/n)^n$ approaches as n becomes ever larger.

Another way to put this is that e dollars would be the amount that \$1 of savings would be worth after a year if the bank compounded its annual 100% interest on it *continuously*. The word continuously here means that you let n run to infinity. Of course, you couldn't actually do that in a real-world calculation. Still, if you wanted to very thoroughly squash the poor schlemiel's hopes of getting rich, you could program a computer to carry out the calculations shown above with a very large n to get an extremely precise approximation of e. Rounded to the nearest hundred-billionth e is 2.71828182846—his dream plainly doesn't compute.

But here's a curious thing about modest little e that sets it apart from bombastic numbers that end in scads of zeros: no matter how long you allow the computer to crank away with ever larger numbers for n, you'll never be able to calculate its exact numerical value. That's because the digits to the right of *e*'s decimal point go on forever in a random pattern—Euler actually established this in 1737. In other words, *e* effectively encapsulates the infinite. (Take that, bazillion, you measly pretender.) You might just as well save yourself time and program the computer to print out "e = 2.71828...and here be monsters."

I find all this lovely and unexpected. Pondering the value of a ridiculously tiny bank account leads directly to one of the grandest conundrums of all time: how to conceptualize infinity without making the brain explode. (Or, as twentieth-century German mathematician Hermann Weyl put it with greater gravitas, the goal of mathematics is "the symbolic comprehension of the infinite with human, that is finite, means.") It's impossible to express *e* in the direct, unabridged way that you can write down, say, 4 or 60,732.89. Basically ignoring this issue, mathematicians made use of *e* and similarly elusive numbers such as π in calculations long before it was known how to define them without dubious hand-waving about their infinite aspect. Rigorously defining *e* requires the concept of a limit, which remained a fuzzy idea in math until the nineteenth century.

ONE OF THE MOST INTRIGUING things about *e* is that it often turns up in calculations that seem unrelated to growth. For instance, it pops out in the solution of the well-known (at least to math lovers) hat-check problem. There are several variations of the problem, which dates from Euler's time. Here's my version: As the guests depart a posh party, a butler hands them their hats, which he'd carefully labeled with their names on sticky notes as they arrived. But since he'd surreptitiously knocked back a few too many glasses of the Château Lafite Rothschild during the party (one of his duties was fetching bottles from the wine cellar), he ignores the stickies and tipsily hands the hats back at random. What's the chance that not a single guest gets the right hat?

It turns out that this probability gets ever closer to 1 divided by e as the number of guests gets ever larger. Using 2.718 as an approximation for e, my calculator shows that $1/e$ equals roughly 0.37, which means there's about a 37% chance that every guest walks out with the wrong hat.

Strangely, the probability is about 37% regardless of whether there are 50 or 50,000 guests. In other words, the chance of the worst-case scenario for hat mix-ups barely changes as the number of guests increases without bound. I don't know about you, but this isn't what I expected when I first encountered the problem.

Compound interest, probability, Euler's formula—these are only a few of many math topics involving e. In fact, e crops up almost as regularly in certain areas of math as Waldo does in *Where's Waldo?*

But e's greatest claim to fame is that when dressed up with a variable exponent,* it becomes a very special **function**. (See box.) This function is usually written as e^x, that is e raised to the x power.

---

*A variable is a proxy for a number that hasn't been pinned down. (In contrast, a specific number like e is called a constant in math.) The appearance of a variable in an equation typically implies a question: What constants can be plugged in for the variable to make the equation true? In $x - 2 = 4$, the variable x is an "unknown" that, when the equation is solved, is shown to be a proxy for a single constant, 6.

Functions are basically variable-containing expressions, such as $x + 5$. (Math books give a more general definition, but this will do for our purposes.) They're like computer programs that convert input numbers into unique output numbers in specified ways, and they're usually designated with equations such as $f(x) = x + 5$, where $f(x)$ means "a function whose variable is x." (Euler came up with this handy $f(x)$ notation, by the way.) When 2 is plugged in for x in $f(x) = x + 5$, the function outputs 7. The math shorthand for that is $f(2) = 2 + 5 = 7$. Another function is $f(x) = 3x^2$. (The expression $3x^2$ means 3 times x^2—multiplication signs are usually omitted between numbers and letters in math, and so when you see a number adjacent to a letter it's safe to assume that there's an implied multiplication sign between them.) For this function, $f(2) = 3 \times 2^2 = 3 \times 4 = 12$, where the ×'s stand for multiplication, not a variable.

If a positive whole number is plugged in for x, e^x means e multiplied times itself x times. For instance, if x is set equal to 2, e^x becomes e^2, or e times e, and since e is about 2.718, that's approximately 7.39. By definition, however, the x in e^x isn't limited to positive integers—you might even plug in a hideous fraction like 197/23 if you were feeling perverse.

Which of course raises a question: How do you multiply a number times itself a non-integer number of times? This might seem as absurd as trying to identify the letter that comes between A and B. But like Lewis Carroll's White Queen, who was capable of believing as many as six impossible things before breakfast, mathematicians are always trying to think their way out of boxes. And beginning in the 1300s, they worked out a perfectly sensible way to expand the definition of exponents to accommodate fractional ones like 197/23.*

I'll skip the minutiae of the expanded definition. But just so you won't feel that the White Queen has stolen a march on you, let's briefly consider how to get a loose grip on a number with a fractional exponent: $2^{5/2}$. Since the exponent, 5/2, or 2½, is more than 2 and less than 3, it stands to reason that $2^{5/2}$ would be more than $2^2 = 4$ and less than $2^3 = 8$—after all, the bigger the exponent attached to a number greater than one, the bigger the result. Thus, $2^{5/2}$ should be equal to a number between 4 and 8. My calculator says $2^{5/2}$ is about 5.66.

Now back to e^x—just why *is* this little function so special? To answer that I'm going to take you on a fast flyover of calculus at 30,000 feet.

CALCULUS IS MUCH CONCERNED with instantaneous rates of change. That's pretty abstract, so here's a real-life example: You're late for a meeting and zipping along the highway, paying little attention to the speed limit. Suddenly

---

*France's Nicholas Oresme, who lived in the fourteenth century, was the first mathematician known to consider fractional exponents; Isaac Newton pinned down their modern meaning in the 1600s. But actually calculating the value of a number with an unwieldy exponent like 197/23 would have been extraordinarily difficult before the invention of logarithms in the seventeenth century. Logarithms make evaluating expressions with bulky exponents ridiculously easy by transforming the calculations, which often seemingly entail gigantic amounts of multiplication, into simple little procedures that take only a few seconds. The great irony about logarithms is that they were invented to make life easy, yet math students often see them as making life hard—which to me is evidence of the fact that they are usually taught with scant attention to how and why they came to be part of mathematics.

you spot a cop up ahead with a radar gun. You desperately start decelerating just before he aims it at your car. What's your speed at the precise moment that he registers it?

This question is more complex than it appears at first glance. In principle, it involves quantifying a rate of change (where what's changing is your car's position) at a single point in time. One difficulty is that it's surprisingly hard to define what's meant by "a point in time." It's natural to think of a particular instant as analogous to a point on the number line. But that leads to problems. For instance, things get sticky when we try to answer this question: Is it possible for one point in time, call it t_2, to *immediately* follow another one, represented by t_1?

Surely the answer is yes—if it were no, how could time go by? But consider: we can identity a point in time between any t_1 and t_2. For instance, we can find a time point between 1:01 and 1:02 p.m. by simply picking the midpoint in time between them, 1:01:30 p.m. Similarly, we can find the midpoint in time between any t_1 and t_2 we choose, even if they're only a gazillionth of a second apart. That suggests there's no such thing as immediately consecutive points in time. In short, there's a paradox hidden inside our intuitions about time.

The root of the problem here is that when we use terms such as "single instant" or "a point in time," we're edging into the baffling realm of the infinite—embedded in the intuitive meaning of such terms is the idea of an infinitely small tick of the clock's second hand. We'll be taking a number of side trips into this clock-melting realm on the way to Euler's formula. As Euler discovered, the way to his celebrated formula leads right through it.

But since the topic at this point is calculus, we'll just skip over the realm lightly. Indeed, calculus could be described as a set of very clever, jack-be-nimble tricks for jumping right over infinity, and the infinitely small, as if they posed problems no more boggling than simple arithmetic does. I'm not going to get into the specifics of the tricks here—this is a high-altitude flyover, and besides, there are many good books and online primers on basic calculus if you're interested. But just for fun let's briefly zoom down for a closer look at the kind of tricky problem that calculus was invented to solve.

Taking up where we left off earlier, picture yourself right after the police officer with the radar gun flags you down. "Jeez, officer," you say when he (the cop could be a she, of course, but I'll stick with he to save letters) walks up beside your car, "I wasn't going any speed at all when you radared me."

"Yeah, right," he might reply. "The radar said you were doing 50. A sign you just passed told you the limit's 30."

"Please let me explain," you reply. "As you doubtless recall from grade-school math, officer, miles per hour means miles divided by hours. But zero hours went by at the instant that you measured my speed. After all, that's what the word instant implies—no time at all goes by at a single point in time. Thus, to calculate my speed at that moment would have required dividing by zero. But dividing by zero is strictly forbidden by the laws of mathematics. It's an undefined operation."

"And so," you conclude with a perfectly straight face, "you can't put a number on my speed at the moment that you supposedly measured it. It would never stand up in court. Judges are very logical, you know, and they have great respect for the laws of math."

After pondering this a little, the cop attacks your key premise: "That's the silliest excuse I've ever heard from a speeder," he growls. "There's nothing wrong with dividing by zero."

"Now hold on," you reply, pulling out pen and paper. "If dividing by zero were allowed, the entire number system as we know it would basically go up in a puff of smoke.

"Here's why: Let's say that it's OK to divide, say, the number one by zero. As you doubtless also recall from grade school, one divided by zero is equivalent to the fraction 1/0." (At this point you start writing numbers as you talk.) "If 1/0 were allowed as a number, then we'd have both $1/0 \times 0 = 1$ (just as $1/2 \times 2 = 1$) and $1/0 \times 0 = 0$ (just as $1/2 \times 0 = 0$). It would follow that one and zero would be equal to the same thing, namely $1/0 \times 0$, and so they themselves would be equal, or $1 = 0$.

"Now pick any number at all, say 50, and multiply it times both sides of $1 = 0$. After this operation, we'll still have two equal numbers, since the two sides of the equation will be changed in an identical way. Thus, we'll have $50 \times 1 = 50 \times 0$, or $50 = 0$. This shows that we can now prove that any number is equal to zero—there's no way around this conclusion if dividing by zero is allowed. And that means the entire number system just went poof!

"Please bear in mind, officer, that when I contest the ticket that you're now so energetically writing, you'll have to explain to the judge that it's no big deal if every number is equal to zero, which, as I've just shown you, logically follows from your starting assumptions about the whole thing. And if the judge buys that, then I'll have to point out that the 50 miles per hour you alleged I was going was actually equal to zero miles per hour. Your own argument will prove that I was just parked by the side of the road when you apparently got all bollixed up fiddling with that radar gun. I always knew those things were dangerous."

PLEASE DON'T THINK that the officer portrayed in this little roadside comedy was unintelligent. It required two great mathematicians—Isaac Newton and Gottfried Wilhelm Leibniz—to figure out how to calculate instantaneous rates of change by inventing calculus. And, as I'll explain later, it took mathematicians another two centuries after that to formulate the techniques involved in such calculations in a really solid, convincing way.

Thus, if the cop had taken calculus (or perhaps even better, read this book), he'd have easily slushed your snow job. In fact, differential calculus (one of the two main branches of calculus—the other is integral calculus) is all about procedures for manipulating functions representing change in order to answer such questions as, "If a car goes $8x^2$ feet in x seconds while peeling out from a stop sign, what is its instantaneous speed after exactly five seconds?"

Unfortunately for math students, applying these procedures to certain functions—actually, to a whole lot of them—is really hairy. This difficulty is arguably the subject's chief source of pain. (It arises in both branches of calculus, by the way, and is especially problematic in integral calculus, which concerns calculating such things as areas and volumes.) And that gets us to what's special about the function e^x: Applying the procedures to it to calculate instantaneous rates of change is laughing-out-loud easy, because they don't change it at all. This means that calculus problems involving e^x are typically no-brainers. For instance, if you're told that the distance traveled by a car as it accelerates away from a stop sign can be expressed as e^x, meaning that it travels e^x feet in x seconds for a brief period after taking off, you immediately know that its instantaneous speed will also be e^x (feet per second at time x) at any moment during that period. No other function has this infinitely user-friendly property.* (Which, fair warning, will be alluded to later.)

What makes this property particularly useful is that a very common form of change can be modeled with functions based on e^x. Called exponential growth (or exponential decay for phenomena that are shrinking), it involves growth or shrinkage at a rate that's proportional to the quantity of whatever it is that's growing or shrinking. An example would be the spreading of a virus that infects people at a rate proportional to the number of people already infected. Other examples: population growth, the radioactive decay of plutonium, and the dissipation of beer froth. (The latter has been empirically confirmed in multiple studies at the college level.)

For our purposes, e^x has an even more special property: it's central to Euler's formula. The formula's first term, $e^{i\pi}$, is just e^x with an unusual, two-part constant plugged in for the variable exponent. This constant is $i\pi$, which consists of a number designated by i times π, the familiar number (pi) having to do with circles. As I'll explain, and as Euler brilliantly proved, plugging in such i-containing numbers for the x in e^x effectively endows the function with the ability to model another common form of change: the oscillatory change of phenomena, such as alternating

* To be accurate, however, I must add that functions of the form c times e^x, where c is a non-zero number, share this property. But these are just minor variations on a theme whose key element is e^x.

current, sound waves, or the back-and-forth motion of a child on a swing. In fact, if I were in sweeping-documentary-film mode, I might summarize the history of Euler's formula as the story of how a great explorer boldly trekked into the mind-bending realm of the infinite to discover that this surprising power lay hidden within a familiar little mathematical expression, and of how later mathematicians, scientists, and engineers used it to help change the world.

CLEARLY *e* IS DIFFERENT from child-safe numbers such as four or 10, which wouldn't dream of inducing sudden loss of cranial integrity. But this wantonness isn't peculiar to *e*. In fact, the number line is chock full of numbers, like *e*, whose decimal representations are effectively infinite. They're called irrational numbers.

Irrationals are numbers that can't be expressed as fractions such as 2/3, 5/2, or 3/1. Another way to put this is that it's impossible to represent an irrational number as the **ratio** of two integers.

Ratios are often expressed in terms like 2 to 3, or 2:3. But they convey the same numerical relationships that fractions do. For instance, if a recipe specified a sugar-to-flour ratio of 1 to 3, it calls for a mixture of the two that is 1/4 sugar and 3/4 flour.

Numbers that can be expressed as such ratios are, sensibly enough, called rational numbers.

There's a little more to this story. All numbers that can be expressed as fractions (the rationals) fall into one of two categories when converted into decimals: repeating or non-repeating. The fraction 1/2 is a non-repeater—its decimal equivalent is 0.5—while 1/3, equal to 0.33333..., where the 3s go on endlessly, is a repeater. (Non-repeaters, by the way, can also be thought of as repeaters ending with infinite strings of zeroes.) Irrationals like *e* fall into neither category—their decimal representations go on forever like 1/3's, but there's no pattern to the numbers after the decimal point. This patternless-going-on-forever aspect of the irrationals means that they can't be completely written out. It also means that every irrational number represents a portal to the infinite.

Although irrationals are regarded as perfectly copacetic today, ancient Greek mathematicians were dismayed when they ran across them about 2,500 years ago. This momentous discovery is credited to the followers of Pythagoras, a mathematician, philosopher, and mystic whose ideas greatly influenced later Greek thinkers such as Plato. The Pythagoreans are much shrouded in the mists of time. Most of the stories that have been handed down about them are just that—legends recorded centuries after they lived. According to one of the stranger tales, Pythagoras had a bean phobia that proved fatal: when, as an elderly man, he was being chased by enemies, he supposedly came to a bean field and refused to enter it, allowing them to overtake him and cut his throat. (You can tell this story is probably untrue, however, because someone as smart as Pythagoras would surely have realized that he could have bought off his pursuers by offering them a piece of his legendary golden thigh.)

The story of most interest here is the one about how the Pythagoreans discovered that certain numbers, such as the length of the diagonal of a square whose sides are one unit long, can't be expressed as fractions. That revealed the existence of the irrationals, and although the Greek mathematicians had only a rudimentary idea about the nature of such numbers, they regarded them as hideously weird. Legend has it that Hippasus, the mathematician who brought attention to the weirdness, was drowned by his fellow Pythagoreans as punishment. His revelation clashed with their religious fixation on positive whole numbers and their ratios, which seemed the perfect building blocks of a well-wrought universe. The existence of numbers that didn't fit into this scheme threatened to blow their whole worldview to smithereens.

The yarn about Hippasus must be taken with enough salt to induce high blood pressure. But it does seem possible that the Pythagoreans were gripped by a kind of dread when they stumbled onto a doorway into the dumbfounding realm of the infinite that existed in the middle of their tidy mental world. And after the irrationals came to mathematicians' attention, it took them well over two millennia to develop the conceptual tongs needed to handle the disturbing numbers in such a way that their explosive implications were kept at a safe remove.*

---

*In 1872, German mathematician Richard Dedekind devised such tongs by defining the irrationals, roughly speaking, as cracks between rational numbers along the number line. More precisely, he characterized irrationals in terms of "cuts" that divide the line's rational-number elements into two sets, with all the numbers in one set, call it A, less than all those in the other set, call it B. If A has no largest element and B has no smallest element, then the pair of sets represents an irrational number. Don't worry if this seems abstruse—the important thing to notice is that it's based solely on rational numbers. (Which, in turn, are made from integers—and what's not to like about high-integrity whole numbers?) It also enables the irrationals to be constructed without reference to the boggling I-thing (∞). In short, it's a rigorous little masterpiece of evasion.

CHAPTER 3

It Even Comes Down the Chimney

The number for the ratio between the circumference and diameter of circles, π, may seem prosaic because it's so familiar. For instance, it is widely celebrated in festivals on March 14 during which people talk mathematics between bites of pie. (That date written as 3/14 shows the first three digits of π, which is approximately 3.14159.) But in truth it's borderline eerie—like e, it seems to pass through the walls that divide mathematics into different subject areas as if they weren't there.

The number π is like e in another important way: it's an irrational number, which was proved in 1761 by Swiss mathematician Johann Lambert, one of Euler's contemporaries. In 1882, π was demonstrated to possess a more unusual property by German mathematician Carl Louis Ferdinand von Lindemann: it's a **transcendental number** (see box), a type of irrational that's

A transcendental number is defined as a number that isn't the solution of any polynomial equation with integer constants times the x's. The term "polynomial equation" refers to the basic kind of equations that algebra students are asked to solve. (Equations, that is, with x's raised to various integer powers and multiplied times constants.) An example of such an equation is $x^2 - 2x - 35 = 0$. Seven is a solution of this equation, meaning that when 7 is plugged in for x, the equation is true. That rules out 7 as a transcendental. We could also rule out 7 as a transcendental by noting that it's a solution of $x - 7 = 0$, $x^3 - 343 = 0$, and an infinite number of other polynomial equations. It's generally easy to prove that numbers encountered in basic math aren't transcendentals by devising polynomial equations for which they're solutions. But proving that a given number is transcendental can be fiendishly difficult. In fact, the only familiar numbers known to be transcendental are π and e. Interestingly, e^π (e raised to the π power) is also known to be transcendental, but no one has been able to determine whether π^e, e^e, and π^π are transcendental or not. The term transcendental refers to the fact that such numbers lie outside (or transcend) the "algebraic" set of numbers that can be solutions of polynomial equations.

extra-far-removed from the integers, fractions, and other relatively ordinary quantities encountered in arithmetic and algebra. (e is also transcendental, which was proved in 1873 by French mathematician Charles Hermite. Mathematicians, including Euler, suggested the existence of transcendental numbers during the seventeenth and eighteenth centuries. But none were actually known to exist until 1844, when French mathematician Joseph Liouville proved that a group of infinitely complicated fractions he'd dreamed up were transcendental.)

The most remarkable thing about π, however, is the way it turns up all over the place in math, including in calculations that seem to have nothing to do with circles. Highlighting this point, physicist Eugene Wigner related a story about a statistician who showed a friend some formulas involving π that are routinely used for analyzing population trends. After noticing π in the equations, the friend exclaimed, "Well, now you are pushing your joke too far. Surely the population has nothing to do with the circumference of a circle."

As nineteenth-century mathematician and logician Augustus De Morgan once mused, "this mysterious 3.14159... comes in at every door and window, and down every chimney." He might have added that once it manages to creep into a room where a mathematician is scribbling away, it likes to sneak onto a page of equations where it seemingly has no right to be and then freeze in place there with a Cheshire Cat-like smile on its face.

In 1671, for example, Scottish mathematician James Gregory discovered an astonishing equation into which π seemed to have quietly slipped while he was playing around with infinite sums. The equation implied that when fractions with consecutive odd-integer denominators are combined in this straightforward way,

$$1 - 1/3 + 1/5 - 1/7 + 1/9 - 1/11 + \cdots$$

(where '$\cdots$' means that the pattern of alternately added and subtracted fractions is continued forever) the grand total equals precisely one-fourth of π, or $\pi/4$. (In math, such infinite combinations of similar fractions are called series. Today mathematicians would say that the limit of this series is $\pi/4$.) Three years later, Leibniz, the German mathematician who co-invented calculus, independently made the same discovery. Historians believe the first discoverer of this astonishing math fact was an Indian mathematician who lived in the fourteenth or fifteenth century.

Proving that this infinite sum of fractions equals $\pi/4$ involves trigonometry, which, as I'll show you later, is much concerned with circles. This suggests that there's a link between the infinite sum and circles, which in turn makes the connection between π and the sum seem at least somewhat plausible. Math is full of such surprising connections, which is one of its greatest attractions. Indeed, mathematics is said to be the perfect subject for conspiracy theorists because when an unexpected connection turns up in it, something is very likely going on that wants explaining. (I've not been able to determine who first made this witty observation, but it bears repeating. However, the mathematicians I know are a lot smarter than the typical conspiracy theorist.)

While the link between Gregory's infinite sum and π can be explained via a fairly involved mathematical proof, it is by no means obvious at first glance. And imagine how you'd feel if you'd lined up the perfectly orderly row of simple, innocent-looking fractions shown above during a math-phobia recovery session and seemingly out of nowhere this infinitely complicated beast of a number jumped up screaming in your face—a transcendental no less, which, by the way, has somehow gotten itself eternally trapped in the structure of circles. You'd be freaked out, and rightly so.

SUCH STRANGENESS IS ANOTHER example of the kind of thing that can happen when you cross over into the zone of the infinite. The portal in this case was that little group of three dots at the end of the line of fractions shown above. During Euler's time, mathematicians were particularly fond of this portal; they developed infinite sums like the Gregory-Leibniz one that, among other things, enabled estimating irrational numbers such as π and e with unprecedented accuracy. On one of his frequent forays into the I-zone, Euler showed that "transcendental functions" such as e^x could be recast as infinite sums. We'll later see how that led to his celebrated formula.

But it's perilously easy to get confused during encounters with the infinite. Consider this question: $1 - 1 + 1 - 1 + 1 - 1 + \cdots =$ what? The answer is obviously 1 if you write the infinite sum this way: $1 + (-1 + 1) + (-1 + 1) + \cdots$, since all the $(-1 + 1)$ terms equal 0. Sticking in parentheses to specify which operations to do first seemingly shouldn't change anything—after all, $2 - 3 + 4$ equals 3, as do both $(2 - 3) + 4$ and $2 + (-3 + 4)$. But the sum appears to be equal to 0 if you write it this way: $(1 - 1) + (1 - 1) + \cdots$, since each of the added terms is obviously 0.

A person who was formerly of sound mind might conclude from this that $1 = 0$ (because both 1 and 0 apparently equal the same infinite sum), which, as noted in the previous chapter, is a detonator for blowing away the entire number system. This thought-twister is called Grandi's series—it's a much-contemplated puzzler in mathematics. Euler believed that its sum equals 1/2, as did other mathematicians of his time. Today, it's considered a "divergent" series, meaning that you can't affix a value to its sum—it endlessly wobbles back and forth between 1 and 0 as you go about adding it up.

Then there's the fact that if you treat infinity like a number and try to do arithmetic with it, you soon find yourself drawing wacky-sounding conclusions like "infinity plus infinity is equal to infinity, and therefore infinity is twice as big as itself." (Zero could inspire the same sort of strange statement, since two zeroes added together are equal to zero. But since zero has no largeness associated with it, there's less temptation to conclude that it is twice as large as itself.) By the same logic, infinity is also a bazillion times as big as itself. By why stop there? The same argument leads to the conclusion that infinity is an *infinite* number of times as big as itself.

You might conclude from this insanity that it's not a good idea to assume that infinity is a number. But asserting that there's no such number seems to imply that the sequence 1, 2, 3,... comes to an end at some point. That would be as counterintuitive as saying, "infinity is an entity that is twice as big as itself."

Perplexity about the infinite reached its apogee in ancient times when Zeno, a pre-Socratic Greek philosopher, originated a series of famous paradoxes concerning it. We've already met him indirectly—the paradox about points in time mentioned in Chapter 2 was a takeoff on one of his befuddling themes. Greek thinkers were basically baffled by the infinite until Aristotle came up with a clever way to think about it that held sway for the next two millennia: infinity, he argued, isn't an actual thing. But there are "potential infinities," like the counting numbers (1, 2, 3,...), that "never give out in our thought."

Aristotle's argument didn't solve all the problems that infinity poses, nor did it end the philosophical fussing about it. But his clever conceptualization, which might be summarized as, "infinity sort of exists, but not really, because it's an imaginary process, not a true thing," let people tiptoe around the infinite without blowing their minds. Potential infinity, and related process-invoking ideas that future mathematicians would use, such as approaching a limit, built on the conceptual foundation that Aristotle laid.

With the advent of calculus in the 1600s, however, a new set of difficulties involving the infinite arose in mathematics. The function-manipulating procedures in calculus that make it possible to quantify instantaneous rates of change were formulated by introducing vanishingly small numbers called infinitesimals into calculations. These speck-like numberlets were thought of as tiny, finite quantities, and yet, when it was convenient to do so in calculations, they were treated just like zeroes. Newton called them vanishing quantities, a rather unfortunate term that highlighted their dubious nature—it made them sound like props used in magic tricks. Anglican bishop and philosopher George Berkeley famously lampooned the implicit contradiction at the heart of the new math of the era by pointing out that infinitesimals were "neither finite quantities nor quantities infinitely small, nor yet nothing. May we not call them the ghosts of departed quantities?"

During the 1800s, mathematicians eradicated the troubling numerical spooks when they rigorously reformulated the foundations of calculus. But in the late nineteenth century, German mathematician Georg Cantor spearheaded a revolutionary way of thinking about infinity that dismissed Aristotle's "it's not really real, folks" approach and forced mathematicians to confront unsettling issues about the infinite once again. Cantor couched his new ideas in the language of set theory, which concerns groupings such as the positive integers, all the fractions between zero and one, and the irrational numbers. He proposed that such sets possess actual, not just potential, infinity.

Cantor happily embraced strange-looking arithmetical statements like "infinity + infinity = infinity." The most boggling implications of his theory, however, concerned the relative sizes of infinite sets—he demonstrated that some such sets are actually larger than others. For instance, he proved that the irrational numbers possess a bigger degree of infinity than the rationals do. In fact, there's an infinite number of levels of infinity, according to Cantor's theory.

This is far-out stuff, and it was dismissed as feverish ravings by a number of Cantor's eminent contemporaries. French mathematician Jules Henri Poincaré, for example, declared that Cantor's theory was a "grave disease" infecting mathematics. But Cantor also had defenders. Perhaps his most prominent advocate was the great German mathematician David Hilbert, who famously asserted in 1926 that "nobody shall ever expel us" from the "paradise" of Cantor's theory of the infinite. Tragically, Cantor, who suffered from recurring bouts of depression, died in a mental hospital eight years before Hilbert wrote that ringing declaration. It isn't clear whether the horrendously tricky challenges he took on contributed to his worsening mental instability. But they may well have done so, at least indirectly, because of all the stress he faced when attacked by prominent mathematicians who dismissed his ideas as ridiculous.

Yet mathematicians have been drawn to infinity through the ages like moths to flames. Or perhaps I should say, like night-hikers to the glimmering of distant fires. Unlike philosophers, who are attracted to the infinite because of its endless fertility as a subject of debate, mathematicians mainly regard infinity as a highly useful conceptual tool (albeit a somewhat tricky-to-use one) for solving practical problems. This is why its symbol, ∞ (sometimes called the lazy eight), which was introduced in 1655 by English mathematician John Wallis, is ubiquitous in calculus and other areas of math. As mentioned earlier, Euler invoked infinity to make a number of his landmark advances, including the derivation of the general equation that led to Euler's formula.

In short, infinity is like a colossal dragon that's known for inducing madness in those who dare to stare hard at it but which is also known for making an honest living by traveling around the countryside and hiring itself out to pull farmers' plows.* (Which is yet another paradox about it.)

BUT LET'S GET BACK to π and its intriguing history. If you're short on time, here's a one-sentence summary: the story of π is the deeply ironic tale of one thinker after another trying to nail down the size of a number that is fundamentally immeasurable. (Because it's irrational.)

The number we now call π has fascinated people for millennia—its study is said to be math's oldest research topic. The main reason, of course, is that it's very handy for calculating

*This simile was suggested by Bill Peet's delightful children's classic, *How Droofus the Dragon Lost His Head*.

the circumferences of circles. For instance, if you wanted to know how long to make a metal band to go around a chariot wheel, you could simply measure the wheel's diameter with a ruler and multiply that length by π.

We don't know who first realized that a number slightly larger than three can be multiplied by the diameter of any circle, no matter the size, to yield its circumference. But knowledge of this fact goes back at least to the time of the ancient Egyptians and Babylonians some 4,000 years ago. Using the Greek letter π to represent the number, however, didn't become standard in math until the mid-1700s, when—who else?—Euler put his imprimatur on that symbol for it.

After it dawned on people that a single number was universally applicable to circle-related calculations, it probably wasn't long before they began trying to express it as a ratio of two integers—a fraction.* Thus began the long quest to pin down the value of π.

Since it's impossible to express an irrational number such as π as a fraction, the quest for a fraction equal to π could never be successful. Ancient mathematicians didn't know that, however. As noted above, it wasn't until the eighteenth century that the irrationality of π was demonstrated. Their labors weren't in vain, though. While enthusiastically pursuing their fundamentally doomed enterprise, they developed a lot of interesting mathematics as well as impressively accurate approximations of π.

---

*Approximating π as a decimal number such as 3.14159 was still far in the future during the time of the ancients. Our modern decimal representation of numbers, which was originated in India and transmitted to the West by Arab mathematicians, was established in Europe during the sixteenth century.

The ancient Greek mathematician, Archimedes, came up with one of the best early approximations by using regular polygons (many-sided figures with equal-length sides, such as stop-sign octagons) that had so many sides they were nearly circular. Calculating a many-sided, regular polygon's perimeter and then dividing it by the length of a diameter-like line through its center yields an approximation of π. Employing polygons with 96 sides, Archimedes showed via this method that π is very slightly less than 22/7, a number that was long mistaken by lesser sages as its exact value.

Chinese mathematician Zu Chongzhi topped Archimedes in the fifth century by approximating π as 355/113—this fraction's decimal equivalent is good to an accuracy of six decimals. It's not clear how he arrived at this remarkably accurate approximation, but historians think he based it on calculations involving a hypothetical polygon with 24,576 sides. In any case, his computations must have been very meticulous, to say the least. It took mathematicians about a millennium to find a more accurate approximation.

During the 1600s, π chasers abandoned the polygon approach in favor of using infinite sums like the one found by Gregory and Leibniz. Mathematicians eventually discovered a splendid array of infinite sums that could be used to calculate approximations of π. Some are far superior to the Gregory-Leibniz formula because fewer of their terms need to be summed to derive a close approximation. One of the most elegant, amazingly simple ones was established by Euler when he was in his late 20s:

$$\pi^2/6 = 1/1^2 + 1/2^2 + 1/3^2 + 1/4^2 + \cdots.$$

This formula is even more shocking than the Gregory-Leibniz one—it reveals a startling connection between π and the positive integers, 1, 2, 3,.... Before Euler derived it, several mathematicians, including Leibniz, had tried unsuccessfully to figure out what all those similar fractions add up to; the problem was first posed in 1644 by Italian mathematician Pietro Mengoli. If you start adding them from the left, you'll soon find that the total sum seems to lie between 1 and 2. If you add up enough of them, you can ascertain that the sum lies in the vicinity of 1.64. But eighteenth-century mathematicians weren't satisfied with such approximations—they wanted to know *exactly* what number the sum equals.

This puzzle, known as the Basel problem (named after the town in Switzerland), was considered one of the most significant questions in math at the time. Thus, when the young Euler demonstrated to everyone's astonishment that the mystery number is exactly $\pi^2/6$, he catapulted himself to international fame. He also provided truly stunning evidence of the uncanny ability of π to sneak through windows and down chimneys.*

---

*Please pause for a moment to ponder these questions: What do circles, implicitly alluded to in the formula by π, have to do with the counting numbers that every school child knows? How do we reconcile π's infinite numerical randomness as an irrational number with the infinite sum's perfectly regular pattern? (π^2 is also an irrational number, as is one-sixth of π^2.) After poring over how Euler solved the Basel problem, I still can't think of satisfying answers to these puzzles—that is, explanations that are intuitively compelling enough to force themselves on me in the way that the truth of $2 + 2 = 4$ does. This isn't to say that I feel truly mystified—Euler's astoundingly clever solution makes a lot of sense. But I find myself experiencing a kind chronic surprise about his result, coupled with the sensation that I must be missing something fundamental about π despite having read a lot about it.

Even before Euler came along, the quest for better approximations of π had ceased to matter for improving real-world calculations and had morphed into a competition for bragging rights. By the early 1600s mathematicians had managed to crank out an approximation of π that was accurate to 35 digits, which is far more than needed for any earthly purpose. With only 39 digits, you could calculate the circumference of the observable universe to within the diameter of a hydrogen atom.

One of the most determined π chasers in the nineteenth century was amateur British mathematician William Shanks, who won minor fame by calculating the first 707 digits of π in 1873. A boarding-school owner with ample spare time to pursue his hobby, he reportedly spent many of his mornings calculating digits of π and his afternoons double-checking his morning's work. After nearly 20 years of work he recorded the 707th digit and moved on to other calculations. Later he proudly commented, "Whether any other mathematician will appear, possessing sufficient leisure, patience, and facility of computation, to calculate the value [of π] to a still greater extent, remains to be seen."

Poor old Bill. In 1944 one of Shanks's successors in the π chase discovered that he'd made a mistake on the 528th digit, which meant that all his later ones were wrong too—more than a fourth of his effort, representing years of work, was wasted. Now he's largely remembered for the goof.

With the help of computers, modern π hunters have taken the quest to truly fantastic levels of accuracy. One of the more notable records was set in 1996 when two extraordinarily clever brothers living in New York City, Gregory and David

Chudnovsky, put together a home-brew supercomputer in their Manhattan apartment and used it to churn out nearly nine billion digits of π. At this writing, the π-obsessed tribe has reportedly managed to pin down several trillion digits. Which goes to show that once you get hooked on something that's infinite, you just can't stop.

CHAPTER 4

The Number Between Being and Not-Being

Some numbers, such as π and e, are distinguished by their utility. Others, such as 5, stand out because they conspicuously correspond to real-world things—raise your hand if you can't think of an example. Yet others are notable for unique properties, such as 6, which is the smallest "perfect number," a positive integer that equals the sum of its positive divisors, not including itself. (For 6, these divisors are 1, 2, and 3.) But the number i is special for a decidedly different sort of reason—it's math's version of the ugly duckling. As we'll see, Euler played a pivotal role in ending its troubled early period, when it was regarded as a grotesque, illegitimate number that insisted on sticking around in math no matter how much it was shunned.

Today it's easy to see the beauty of i, thanks, among other things, to its prominence in mathematics' most beautiful equation. Thus, it may seem strange that it was once regarded as akin to a small waddling gargoyle. Indeed, the simplicity of its definition suggests unpretentious elegance: *i* is just the **square root** of –1. But as with many definitions in mathematics, *i*'s is fraught with provocative implications, and the ones that made it a star in mathematics weren't apparent until long after it first came on the scene.

The square root of a number, call it *x*, is a number that when multiplied times itself equals *x*. For example, the square root of 4 is 2. To be precise, 2 is the "principal" square root of 4, and there's also another one: –2, which reflects the fact that a negative number times a negative number is positive.

One implication is that the things about *i* that once vexed mathematicians aren't limited to a single number. That's because *i*, in effect, is the progenitor of an infinity of similar (former) vexers—the imaginary numbers. Each is the counterpart of a **real number** (see box). For example, *i*, or $1 \times i$, is the imaginary-number counterpart of the real number 1, and $-i$, or $-1 \times i$, is the imaginary-number version of –1. If you added four *i*'s together, you'd have $i + i + i + i$, or $4 \times i$, usually written $4i$; this number, of course, is the imaginary counterpart of 4. (You could call it "four times the square root of –1," but it's easier to use the symbol *i*—$4i$ is spoken "four eye.")

All the imaginary numbers have the same form as $4i$—each consists of a real number times *i* to form an imaginary-number version of the real number. An imaginary number of special interest in this book is π times *i*, which can be written as

A real number is a number that lies along the familiar number line. Thus, the reals include positive and negative integers, zero, fractions (a.k.a. rational numbers, which include the integers), irrational numbers, and that intriguingly named subset of the irrationals, the transcendental numbers.

either πi or $i\pi$—the imaginary counterpart of π, it's the exponent in Euler's formula.

But just what are imaginary numbers, you may now be asking yourself, and what on earth could it mean to raise *e* to an imaginary-number power? This chapter concerns mathematicians' long struggle to answer the first of these two questions. Later we'll take up the second one, which inspired Euler to devise the most radical expansion of the concept of exponents in math history. At this point, suffice it to say that affixing an imaginary exponent to a number has a dramatic effect on it—something like what happens to a frog when it's tapped by a standard-issue magic wand.

Before the eighteenth century, the imaginary numbers seemed to have cornered the market on mathematical ugliness. Just doing basic arithmetic with them required right-thinking people to undergo "mental tortures," claimed Italian mathematician Gerolamo Cardano. That's because the term "square root of negative one" was generally thought at the time to be close to perverse gobbledygook. For one thing, the square root of –1 doesn't correspond to a familiar, real-world quantity in the way that, say, 2 does to a couple of flagons of ale, or that 4 does to the area of a square whose sides are two feet long. Negative numbers posed some of the same quandaries that the imaginary numbers did to Renaissance mathematicians—they didn't seem to correspond to quantities associated with physical objects or geometrical figures. But they proved less conceptually challenging than the imaginaries. For instance, negative numbers can be thought of as monetary debts, providing a readily grasped way to make sense of them.

Another disconcerting thing about i and its ilk, however, is that they don't conform to the familiar rules of arithmetic. If you multiply a positive number, such as 3, times itself—that is, if you square it—the result is always a positive number, in this case, positive 9. Same with negative 3, or -3. Square it and you get positive 9. But when you square the square root of -1, by definition you must get a negative number. That's because i is defined as the square root of -1, and so when you multiply it times itself you absolutely must wind up with -1. To put that more succinctly, $i^2 = -1$.

This kind of peculiar arithmetic occurs with all the imaginaries. For example, $(4i)^2$, which is shorthand for $(4 \times i) \times (4 \times i)$—just to be clear, the ×'s stand for multiplication here—must be equal to (by rearranging terms via the commutative and associative laws of multiplication) $4 \times 4 \times i \times i$, or $16 \times i^2$, or 16×-1, which is -16. (The commutative law states that $a \times b = b \times a$, which means that order doesn't matter when you multiply two numbers. For instance, the law implies that πi is the same number as $i\pi$. The associative law specifies that $(a \times b) \times c = a \times (b \times c)$, which means that it doesn't matter how you group multiplied numbers, or which two you multiply together first.)

The fact that multiplying *positive* $4i$ times *positive* $4i$ yields *negative* 16 seems like saying that the friend of my friend is my enemy.* Which in turn suggests that bad things would happen if *i* and its offspring were granted citizenship in the number world. Unlike real numbers, which always feel friendly toward the friends of their friends, the *i*-things would plainly be subject to insane fits of jealousy, causing them to treat numbers that cozy up to their friends as threats. That might cause a general breakdown of numerical civility.

Of course, mathematicians don't worry about discord among numbers. (In fact, I have to admit that my tossing out the idea of such discord represents giddy metaphorical overreach.) But it's not hard to formulate the issue underlying the weirdness here in the kind of terms mathematicians do care about. For instance, assuming that *i* really is a number, let's try to figure out where it should go on the number line—that seemingly all-inclusive home for numbers. (Even the outlandish transcendentals live on it.) We can't position *i* among the negative numbers to the left of zero, because all those numbers, when multiplied times themselves, yield positive numbers. It can't be zero in disguise, because zero times zero equals zero, not –1. And it can't lie among the positive numbers to the right of zero, because they also yield positive numbers when multiplied by themselves. Thus, assuming that *i* and its kind really are numbers seems to entail believing in the existence of a new, *ad hoc* number line filled with peculiar,

---

*I'm referring here to a trick that grade-school students are sometimes taught to help them remember whether multiplying negative numbers times other numbers, for instance –2 × 3, yields positive or negative results: you think of negative numbers as enemies, and positive numbers as friends. Thus, multiplying a negative number, such as –2, times a positive one, such as 3, is pictured as "the enemy of my friend," who of course is my enemy too, or a negative number. It follows that –2 times 3 equals –6.

number-like entities that are totally useless for counting things or measuring them. Why bother?

To be sure, pre-eighteenth-century mathematicians knew that the imaginaries were bound to come up as solutions to certain algebra problems.* But for a long time, such problems were simply dismissed as having no solutions, which implied that the imaginaries had no place in mathematics. They just weren't the kind of sound, yeomanly numbers that the mathematicians of yore could recommend in good conscience to people for counting and measuring.

But the strange numerical aliens kept turning up and pestering math's immigration officials. Most notably, they butted into the solutions of so-called cubic equations in a way that couldn't be ignored.

Cubic equations are ones that include a variable, such as x, raised to the third power—that is, having an exponent of three—but no variables raised to higher powers. An example would be $x^3 - 15x - 4 = 0$. Solving for x in such equations is often quite difficult—actually, it was frequently unmanageable before the sixteenth century—and Renaissance mathematicians regarded finding a surefire way to solve cubic equations as one of the great challenges of the age.

In the 1500s, Cardano and other Italian mathematicians cracked the problem by inventing an ingenious algorithm—a step-by-step procedure like a recipe—to crank out cubic-equation solutions based on the constants in the equations. How-

*One such problem is to solve for x in this equation: $x^2 + 1 = 0$. You might be tempted to try 1 for x, but that doesn't work because $1^2 + 1 = (1 \times 1) + 1 = 1 + 1 = 2$, and 2 is decidedly not 0. Subbing in -1 for x doesn't work either: $(-1)^2 + 1 = (-1 \times -1) + 1 = 1 + 1 = 2$. But i works like a charm: $i^2 + 1 = -1 + 1 = 0$.

ever, the algorithm was a bit dubious because it often produced solutions that had imaginary numbers embedded within them. (Cubic equations typically have three different solutions, meaning that they're true for three different values of x, and it's often the case that two of the solutions involve imaginary numbers.) At first these strange-looking solutions were regarded as worthless and basically fed to the pigs.

But another Italian mathematician, Rafael Bombelli, discovered a cool but disquieting thing around 1570. Instead of discarding a cubic-equation solution that contained imaginary numbers, he played around with it using standard algebraic techniques and showed that it was actually a camouflaged real number.

The equation, in case you're interested, was the one mentioned above, $x^3 - 15x - 4 = 0$, and the camouflaged real-number solution was 4. (The solution is still 4 after all this time, as you can readily check by plugging that number in for x.) The complicated expression that he discovered to everyone's surprise was equal to 4 prominently included the square root of –121, which is equivalent to the square root of 121 (or 11) times i.

Bombelli's discovery showed that it was necessary to treat apparently meaningless imaginary-number-based solutions as legitimate numbers in order to find such hidden real-number solutions. That meant the imaginaries could no longer be cavalierly pig-troughed. But mathematicians still didn't feel comfortable around them. Indeed, with their cubic-equation solver in hand they were like pioneering clockmakers who had figured out how to construct a clock that kept perfect time. But apparently it was haunted, because sometimes instead of chiming at the top of the hour it would issue ghastly screams. People had been in the habit of covering their ears when that happened until a particularly clever clockmaker had discovered that the number of screams indicated the hour and thus could be used to tell time even when the clock was stashed down in the dungeon. But some people still found it creepy to listen to the thing when it was screaming.

Eventually mathematicians got more comfortable with the imaginaries despite continuing to think of them as strange. In 1702, Leibniz blithely observed that "imaginary numbers are a fine and wonderful resource of the divine intellect, almost an amphibian between being and non-being."

A few decades later, Euler effectively granted numerical citizenship to the imaginary numbers by pointing out that "nothing prevents us from...employing them in calculation." In fact, he employed them with boundless enthusiasm—Euler's formula is one of the results—and he encouraged their use by assigning the handy symbol i to the square root of -1, which has been used ever since.

Still, even Euler had trouble pinning down their basic nature, because, as mentioned earlier, they don't correspond to physical quantities or readily envisioned geometrical objects as more familiar numbers do. Essentially throwing up his hands, he wrote that they're "impossible" numbers, and exist "merely in the imagination."

The imaginaries finally lost their air of impossibility when nineteenth-century mathematicians realized that they're actually perfectly ordinary, law-abiding numerical beings—it's just that they hail from a different dimension. We'll go there later on.

CHAPTER 5

Portrait of the Master

The fact that Euler could effectively grant citizenship to an entire class of troubling numbers thought to resemble ghostly salamanders gives an indication of his great sway in math. He even turned the imaginaries into fetching little playthings. What made him so influential? (True, he was a genius, but there have been many math geniuses in history who weren't as important as he was—and still is—in mathematics.) Let's briefly veer off our excursion's main path to get a better look at the one who blazed it.

My favorite description of Euler was offered by Dieudonné Thiébault, a French linguist who met him in mid-life: "A child on his knees, a cat on his back, that's how he wrote his immortal works." Euler loved accompanying his children to marionette shows, where, it is said, he laughed robustly along with the kids at the puppets' antics. He liked joking around with his children and grandchildren and teaching them about math and science. He also liked to take them to the zoo, where he gravitated to the bears—he loved watching the cubs play. He enjoyed having visitors drop by to talk about anything under the sun and was adroit at shifting from deep technical discussions to casual conversation as called for by the occasion. I suspect that cats generally purred in his presence.

Born in 1707 to a Swiss pastor and his wife, Euler seemed likely to follow his father into theology until he got hooked on math as a teenager at the University of Basel. It was a mediocre school overall, but luckily one of the world's greatest mathematician at the time, Johann Bernoulli, taught there. After Euler's talent came to Bernoulli's attention, he took the youth under his mighty wing, giving him special tutorials every Saturday afternoon. Bernoulli assigned Euler increasingly difficult problems to work out by himself, reserving the Saturday sessions to help his student with the ones he had trouble with. But after a while, as mathematician William Dunham has noted, "it was Bernoulli who more and more seemed to become the pupil." A few years after the senior mathematician began mentoring Euler, he was addressing his young protégé in letters with a Latin phrase translated as "The Most Famous and Learned Man of Mathematics." It should be noted that Bernoulli wasn't a humble man, nor was he given to jests.

Euler gained public recognition at age 19 when he first entered the international contest held annually by the Paris Academy of Sciences. That year the challenge was to determine the best placement of masts on ships to capture the wind's pushing power. Euler's submission tied for second place—not bad for a teenager competing against the top mathematicians and scientists of Europe. (And the Swiss youngster had never even seen a large sailing ship.)

Leonhard Euler

As broad as he was deep, Euler made seminal advances in just about every area of math—number theory, calculus, geometry, probability, you name it. He "seems to have carried in his head the whole of the mathematics of his day," marveled twentieth-century French mathematician André Weil. He also put new areas of math on the map.

The breadth partly reflected his prodigious memory. Even in old age Euler could readily recite the 9,500-plus lines of Virgil's *Aeneid* from memory. He knew five languages: Latin, Russian, German, French, and English. It's said that he could reel off the first six positive integer powers of any number between 1 and 100. (In case you'd like to master that trick yourself, here's a start on the 600 numbers you'd need to know: $99^1=99$; $99^2=9,801$; $99^3=970,299$; $99^4=96,059,601$; $99^5=9,509,900,499$; and $99^6=941,480,149,401$.)

Euler apparently could write a groundbreaking mathematics paper in the half hour between the first and second calls to dinner, according to historian Eric Temple Bell. He could do that, Bell added, because he possessed "all but supernatural insight into apparently unrelated formulas that reveal hidden trails leading from one territory to another" in math.

To be fair, it was somewhat easier for Euler to be prolific than it was for the later practitioners of mathematics—they were compelled to be more painstaking than he was as the bar was raised on rigor, and math's branches bore less low-hanging fruit after the eighteenth century's rich harvest. Still, when recently going over Euler's derivations of several landmark equations, including his famous formula, I found myself thinking that, like the imaginary numbers, he must have arrived here from a different dimension—as Bell said, he had a truly uncanny knack for sensing the presence of hidden trails. (Or, as twentieth-century mathematician Mark Kac once put it, "An ordinary genius is a fellow that you and I would be just as good as, if we were only many times better. There is no mystery as to how his mind works. Once we understand what he has done, we feel certain that we, too, could have done it. It is different with the magicians...the working of their minds is for all intents and purposes incomprehensible.")

This book can present only a superficial glance at Euler's overall achievements. But let me offer a sports analogy that occurred to me as I perused a tiny fraction of his vast output.

The math game in Euler's day resembled track and field in the early twentieth century, a freewheeling era for the sport that was memorably portrayed in the Oscar-winning movie *Chariots of Fire*. A champion runner at the time might have strolled up to the starting line of a race puffing a cigar, nonchalantly set it down by the track, taken off like a shot at the gun to handily win, then picked up his still-smoldering stogie and sauntered off to the locker room. It was obviously easier to set records in those days, just as it was to make math advances in the eighteenth century.

But if Euler had been an early track star, he wouldn't have just won a race now and then. He'd have regularly trotted into track meets with a child in his arms and a cat reposing on his back, and, without putting them down, proceeded to win the discus, the hammer throw, the shot put, the javelin, the long jump, the high jump, the triple jump, the steeplechase, the 100-meter dash, the 200, the 400, the 800, the 1,500, and the mile. The kid and cat would have been napping by then, but because he always did his thing with incredible gusto he'd have gone on to set a world record in the 5,000-meter race running backwards in bedroom slippers with a blindfold, and then finished up by topping his own previous world record in the pole vault—somehow without waking the child or the cat.

Euler is said to be the greatest math explainer ever. He authored a classic guide to algebra that has been called the second most popular math book in history after Euclid's *Elements*. (The latter, by the way, is thought to be second only to the Bible as the most frequently printed book.) Euler also wrote widely-used textbooks on calculus and the laws of motion, and a primer on science, philosophy, music theory, and other topics that became an eighteenth-century bestseller: *Letters of Euler on Different Subjects in Natural Philosophy Addressed to a German Princess*. Dedicated to the Princess of

Anhalt Dessau, the niece of Prussia's King Frederick II, the book made Euler something of a pioneer in supporting the education of women on technical topics. (It consisted of letters because Euler had acted as the princess's remote tutor when the Prussian royal court had fled Berlin during the Seven Years' War.) The *Letters* explained such things as why it is cold on top of mountains in the tropics, why the moon looks larger when it's near the horizon, and why the sky is blue. Euler wrote it in French but it was soon translated into all the other major languages of Europe and widely used to teach basic science. The book got rave reviews from, among others, Kant, Goethe, and Schopenhauer.

Euler suffered an infection that blinded his right eye when he was in his 20s, and later a failed cataract operation in his left eye rendered him unable to make out people's faces and even nearby objects. This great loss didn't slow him down a bit. In fact, he cheerfully talked of losing his eyesight as "one fewer distraction." With the help of assistants, he produced more than half of his entire life's work during the last 17 years of his life, after he had lost most of his vision. Ever resourceful, he found a way to get exercise by repeatedly walking around a large round table in his study while running his hand along the edge.

He overcame a number of setbacks and personal tragedies besides going blind. Only five of the 13 children he had with his wife Katharina survived to adulthood, and only three, all sons, outlived him. His house burned down when he was 64, destroying his library and some of his unpublished work. Euler, who was by then nearly blind, was stranded on the second floor during the fire until his Swiss handyman, one

Peter Grimm, heroically climbed a ladder and brought him down over his shoulder. Katharina, his wife of 40 years, died when he was 66; three years later he married Katharina's widowed half-sister so that he wouldn't be totally dependent on his children.

After getting established at the St. Petersburg Academy of Sciences in Russia early in his career, Euler was basically forced out as a wave of anti-foreigner sentiment swept through Russia. It hadn't helped that the main interest of the academy's effective director, Johann Schumacher, "lay in the suppression of talent wherever it might rear its inconvenient head," as historian Clifford Truesdell sardonically put it.

Capitalizing on the situation, Prussia's King Frederick hired Euler to help bolster the Berlin Academy of Sciences. But Frederick never regarded Euler, a quiet, pious family man, as the kind of drawing-room wit he wanted for the academy. The king, despite his pretensions to intellectual breadth, apparently suffered from incurable math phobia. Mathematics "dries up the mind," he once wrote in a letter. Frederick was deeply annoyed whenever Euler attended the theater in his presence—the mathematician was known for getting noticeably distracted from the play as he jotted notes about the hall's optics, sound effects, and other features that could be mathematically modeled.

As the years passed, Euler became the butt of bumpkin jokes made by the academy's favored ornaments, such as Voltaire, whom Frederick had lured from France by offering to pay him some 20 times as much as he'd proposed paying Euler as a starting wage. Since the mathematician's right eye was blind, Frederick smirkingly referred to him in a letter to

Voltaire as "our great Cyclops." In the same letter the king joked that he would be willing to trade Euler to Voltaire's consort, Émilie du Châtelet, in return for Voltaire. In a letter to his brother, Frederick observed that people like Euler are "useful...but otherwise are anything but brilliant. They are used as are the Dorian [sic] columns in architecture. They belong to the subfloor, as support...."

Euler actually did serve as a crucial part of the Berlin academy's support system for a quarter century. He supervised its observatory and botanical gardens, oversaw its finances, managed the publication of its calendars and maps (the academy's main source of income), advised the government on state lotteries, insurance, pensions, and artillery, and even oversaw work on hydraulic pipes at Frederick's summer residence. But finally he got fed up with being treated with contempt and petitioned Frederick for permission to resign from the Berlin academy. At first the king refused to even acknowledge the request, but Euler stubbornly persisted and was set free at last when the king sent him a curt, cold note: "With reference to your letter of 30 April, I permit you to quit in order to go to Russia." Frederick apparently never realized that he had driven away one of history's greatest minds. After later filling Euler's position at the Berlin academy, he commented in a letter that "the one-eyed monster has been replaced by another who has both eyes." (The two-eyed monster was the great French-Italian mathematician Joseph-Louis Lagrange.)

Euler returned in 1766 at age 59 to Russia's St. Petersburg Academy, where the anti-foreigner movement had faded during the reign of Catherine II, and spent the rest of his incredibly productive life there.

EULER IS REGARDED as having had only three equals in the history of mathematics: Archimedes, Isaac Newton, and Carl Friedrich Gauss. I find it interesting to compare their personal characteristics with his. Such comparisons usually say little to nothing about great creators' achievements. But this is an exception. In my view, Euler's tranquil temperament, fairness, and generosity were integral to his greatness as a mathematician and scientist—he was never inclined to waste time and energy engaging in petty one-upmanship (like his mentor, Johann Bernoulli, who was known for getting into the eighteenth-century version of flame wars with his older brother, mathematician Jakob Bernoulli, and even with his own son, Daniel, over technical disputes), brooding about challenges to his authority (like Newton), or refusing to publish important findings because of the fear that they might be disputed (like Gauss).

If the stories told about Archimedes are true, he was a colorful character. It is said that he once jumped from his bath and ran naked through the streets shouting "Eureka!" when he realized how to measure volumes of irregular solids by submerging them in water. According to Plutarch, when Roman soldiers overran Syracuse, where Archimedes lived, the mathematician told one of them he had to finish some calculations before he could meet the conquering Roman general. The enraged soldier drew his sword and killed the ancient world's most brilliant man on the spot.

Newton was a shy, prickly loner who held grudges. He was given to outbreaks of near-psychotic rage when challenged or contradicted. When compiling a list of his sins at age 19, he noted that one of them was "threatening my [step] father and mother Smith to burn them and the house over them." William Whiston, who assisted Newton and then succeeded him as Lucasian Professor of Mathematics at the University of Cambridge, reported that "Newton was of the most fearful, cautious and suspicious temper that I ever knew."

He was notably tyrannical as president of Britain's Royal Society—contradicting his opinions or directives simply wasn't allowed (which is an odd position to take for a scientist).

Newton conceived an especially passionate hatred for the scientist Robert Hooke, who challenged Newton's ideas on the nature of light. Some years later an exchange of letters between the two helped inspire Newton's breakthrough ideas about gravity and planetary motion. After Hooke later suggested that he'd had a role in bringing forth the famous concepts, Newton furiously deleted every reference to Hooke in his forthcoming *magnum opus*, *Philosophiae Naturalis Principia Mathematica* (Mathematical Principles of Natural Philosophy). As historian Robert A. Hatch put it, Newton's "hatred for Hooke was consumptive."

Newton was even more consumed by his famous dispute with Leibniz about which one of them deserved credit for inventing calculus. Newton first developed the basic ideas of calculus, but Leibniz came up with them independently and was the first to publish them. While Newton pretended to be above the fray, he secretly oversaw attacks by his English allies on Leibniz and, behind the scenes, he dominated a supposedly impartial investigation by a committee of the Royal Society that was convened to decide on the priority dispute. The committee didn't bother to give Leibniz a hearing before pronouncing in favor of Newton, who wrote its report himself. To top it off, he then tried to ensure that the report would be widely noticed by writing an anonymous review of it for the *Philosophical Transactions of the Royal Society.* Because of the dispute, English mathematicians spent the next century chauvinistically ignoring math advances on the Continent, where Leibniz and his main allies were, thereby losing their innovative edge in mathematics.

Gauss was similarly forbidding. He disliked teaching, had few friends, and alienated one of his sons, Eugene, by trying to control his life. Gifted with languages, Eugene had wanted to study philology as a youth, a choice opposed by his father. After they quarreled about a dinner party that Eugene had held with his friends and had asked his father to pay for, the 19-year-old son abruptly left for America and never returned. In the United States, he learned the Sioux language and wound up working for a fur company in the Midwest.

Gauss withheld many of his results because he felt they weren't perfect enough to release. But after other mathematicians independently discovered and published the same findings, he wasn't averse to letting people know that he'd privately made them first. One such case concerned the Hungarian mathematician Wolfgang Farkas Bolyai and his son, János. The elder Bolyai was a friend of Gauss, and in 1816 he asked the famous German mathematician if he would let János, then 14, live with him and become his student. Bolyai couldn't afford to send his gifted son to a prestigious university, and Gauss could have been a great help to him. But Gauss refused.

Nevertheless, János went on to help pioneer "non-Euclidean" geometry while he was in his early 20s. A century later this revolutionary development in mathematics would inform Einstein's General Theory of Relativity, which suggests that space-time is curved. While working out the new ideas, the hopeful young man excitedly wrote his father, "I myself have made such wonderful discoveries that I am myself lost in astonishment." A paper he authored on the findings was published as an appendix to a math book by his father that appeared in the early 1830s. Clearly proud of his son's work, Wolfgang sent a copy to Gauss.

Gauss replied that to praise János's work "would be to praise myself"—meaning that he had already discovered everything that János had worked out. That was a severe blow to János. Soon after, his mental and physical health deteriorated. While continuing to work fitfully on mathematics, he never lived up to his early promise and eventually gave up trying to win recognition in math. He died in obscurity, and his innovative work was ignored until after Gauss's death in 1855; the posthumous publication of Gauss's private notes and letters on non-Euclidean geometry made the topic seem worth studying to mathematicians, who then recognized Bolyai's contributions.

Gauss wasn't always so dismissive of young talent. For several years during the early 1800s, he sent encouraging letters to France's Marie-Sophie Germain, one of the first prominent female mathematicians. As a young teen she'd developed a passion for math but was discouraged from studying it by her parents—that was considered unladylike. They'd even tried to discourage her from surreptitiously working math problems during cold nights by denying her warm clothes and a bedroom fire. But they'd finally relented after finding her asleep at her desk one morning next to a frozen inkwell and a slate full of calculations. Although she was barred from attending France's leading school for aspiring mathematicians, she went on to do pioneering work in number theory and other areas.*

Gauss never met her, but in one of his letters to her, he sympathetically (and accurately) wrote that when "a woman, because of her sex, our customs and prejudices, encounters infinitely more obstacles than men in familiarizing herself with [number theory's] knotty problems, yet overcomes these fetters and penetrates that which is most hidden, she doubtless has the most noble courage, extraordinary talent, and superior genius."

It seems that Gauss himself didn't show much in the way of noble courage. In 1829, for example, he confessed in a letter to a friend that he'd long withheld his results on non-Euclidean geometry because he'd feared getting attacked for espousing radical new ideas. And whenever he did disclose his indisputably brilliant discoveries, he often put them in a form that made it difficult for others to follow his thought processes. "He reworked his mathematical proofs to such an extent that the path whereby he had obtained his results was all but obliterated," remarked British mathematician Ian Stewart. Gauss's contemporaries less diplomatically described his writing style as "thin gruel."

---

*Given all the doors closed to women through the ages, it is striking (and hugely instructive) that accomplished female mathematicians have not been terribly rare. A short list would include Hypatia of Alexandria, Maria Gaetana Agnesi, Sofia Kovalevskaya, Alice Boole Stott, Julia Robinson, Emmy Noether, and Mary Lucy Cartwright.

In contrast, Euler was about the nicest, most forthcoming person you could hope to meet in the annals of genius. Cats and children weren't the only ones who gravitated to him—the record shows that almost everybody who met him found him charming. The main exceptions were King Frederick and the people who thought his cyclops joke was funny. Some years after Euler left Prussia, however, he and Frederick had a friendly exchange of letters when the king got interested in a tract by the mathematician on how to set up and calculate pensions—Euler seemed incapable of holding grudges. This isn't to say that he was always mild-mannered. He held strong opinions and wasn't afraid to contradict peers when he thought they were wrong. But his disputes with them were generally cordial arguments rather than bitter fights.

Not surprisingly, the great explainer had a passion for teaching and, according to one story, taught elementary algebra to his tailor by employing him as a scribe when dictating his famous algebra text. Euler's son, Johann, later remarked that the man was capable of solving complicated algebra problems after the experience.

One day in 1763, a young Swiss man, Christoph Jezler, arrived at Euler's house in Berlin. Jezler explained that he wished to copy Euler's yet-to-be-published textbook on integral calculus, page by page. It turned out that as a youngster he'd aspired to be a mathematician, but family pressures had forced him to become a furrier, a trade he'd recently put aside after his father had died. Euler not only took Jezler in but also offered to help him understand parts of the text that he had trouble with. After some months as Euler's guest, during which Jezler furiously copied away while his mother sent cherries, apple wedges, and plums to the Euler household (Euler especially loved the plums), the young man returned home and later became a professor of physics and mathematics.

Euler carefully cited other mathematicians' contributions in papers he wrote related to their work. Sometimes he gave others more credit than they deserved. Once he generously put aside his rapidly advancing work on hydrodynamics, a branch of physics dealing with fluids, so that he wouldn't risk upstaging a friend—mathematician Daniel Bernoulli, the son of Euler's former mentor—who he knew was laboring on a major book about the topic. (Euler also probably wanted to steer clear of a messy dispute between the younger and older Bernoulli concerning which one of them first came up with key ideas on the topic. Daniel's imbroglio with his irascible father, who took credit for much of Daniel's work, made the son thoroughly miserable, and at one point he commented that he wished he'd become a cobbler instead of a mathematician.)

Another time, Euler translated a book on ballistics by England's Benjamin Robins, who held eccentric views on math and physics that had earlier prompted him to mount a ludicrous public attack on a related work by Euler. Euler made the translation much better than the original, adding commentaries and corrections that turned the book into a major work as well as making it much longer than it originally was. As a Euler admirer later noted, the only revenge the great mathematician took on Robins for attacking him was to make the Englishman's book famous.

Euler's even-tempered generosity even extended to those with whom he differed on religion. In 1773, the French philosopher Denis Diderot visited St. Petersburg for several months at the invitation of Catherine II. Diderot was best known as an editor of the *Encyclopédie*, a pioneering, 28-volume encyclopedia that covered everything from mathematics to music to medicine. But it was most famous for advocating Enlightenment ideals, including reason-based challenges to religious dogma.

Euler was a devout Protestant who held daily prayers in his home, which seemingly would have made for conflict between him and Diderot. Indeed, according to a story widely circulated in Europe after Euler's death, the mathematician embarrassed the visiting Frenchman in a public debate about God's existence by exclaiming to Diderot, "Sir, $a+b^n/n = x$, therefore God exists—Respond!" Diderot, who supposedly knew little about math, stood astonished by this absurdity as onlookers burst into laughter. Deeply mortified, he soon after returned to France.

There's no evidence at all that this story is true, according to historian Ronald Calinger. It appears to have been invented by Prussia's King Frederick, or a member of his court, to belittle Diderot, who had incensed the king by publicly criticizing some of his military policies. In fact, soon after Diderot had arrived in St. Petersburg, Euler had arranged for him to be inducted into the Russian Academy of Sciences as a "corresponding member" and had subsequently presided over the induction ceremony. Diderot had responded by sending a letter to the academy declaring that "everything that I created I would gladly trade for a page of the writings of Monsieur Euler." Contrary to the scurrilous story, Diderot was well-versed in mathematics and apparently regarded Euler as one of the leading lights of the age. And for his part, Euler was much too gracious to publicly attack the visiting scholar.

The most-repeated quote on Euler is attributed to French mathematician Pierre-Simon Laplace, who advised fellow mathematicians to "read Euler, read Euler. He is the master of us all." Gauss commented that "studying the works of Euler remains the best school in the various fields of mathematics and cannot be substituted for anything else." Twentieth-century Swiss mathematician and philosopher Andreas Speiser had this to say: "If one considers the intellectual panorama open to Euler, and the continual success of his work, he must have been the happiest of all mortals, because *nobody has ever experienced anything like that*." (Emphasis added to highlight that this extraordinary statement is probably not an exaggeration.) To put that another way, Euler wasn't at the cutting edge—he *was* the cutting edge, one unlike any we've seen before or since thanks to the fortuitous combination of his unsurpassed genius and the immense opportunities for progress opened up by the mathematicians and scientists who preceded him. Given that, it is only to be expected that Euler's formula is deep and strange. In fact, it's a lot like Euler's mind.

CHAPTER 6

Through the Wormhole

If, like Euler, you were given to plumbing depths that no one even realized were there before you revealed them, where better to look for hidden passages to the altogether elsewhere than *e* and π, which harbor a form of infinity and, even more intriguing, were cropping up in mathematics with striking regularity during the eighteenth century? The number *i* must have exerted the same pull on a mind like Euler's. Indeed, $e^{i\pi}$ seems to me exactly like the kind of bizarre, but interesting, expression that Euler would have mused about as he expanded the universe of mathematical ideas—it's positively Euler-esque.

But what about the other two numbers in $e^{i\pi} + 1 = 0$? At first glance, 1 and 0 seem to lack the endless charisma of the equation's other three numbers. But, again, looks can be deceiving in mathematics—these two are also Very Important Numbers. In fact, they're arguably the only numbers that outrank *e*, *i*, and π in the VIN pantheon.

One, in a nutshell, is just that—the first quantity that came on the scene when people starting counting things. It's the mother of all the positive integers: you can generate them from one with the help of addition. It also has a marvelously light touch—one is the one and only number that when multiplied times other numbers leaves them just the way they were. When author Alex Bellos surveyed people's favorite numbers and adjectives associated with them, he was informed that one is independent, strong, honest, brave, straightforward, pioneering, and lonely.

Zero seems as diaphanous as a fairy's wing, yet it is as powerful as a black hole. The obverse of infinity, it's enthroned at the center of the number line—at least as the line is usually drawn—making it a natural center of attention. It has no effect when added to other numbers, as if it were no more substantial than a fleeting thought. But when multiplied times other numbers it seems to exert uncanny power, inexorably sucking them in and making them vanish into itself at the center of things. If you're into stark simplicity, you can express any number (that is, any number that's capable of being written out) with the use of zero and just one other number, one. (The trick for doing that is to use the binary number system, in which numbers are expressed in terms of 1s and 0s.)

However, in contrast to one, which is singularly straightforward, zero is secretly peculiar. If you pierce the obscuring haze of familiarity around it, you'll see that it is a quantitative entity that, curiously, is really the absence of quantity. It took people a long time to get their minds around that. In fact, zero was just a big nothing, conceptually speaking, until Indian mathematicians accepted it as a legitimate number somewhere between the fifth and ninth centuries—thousands of years after numbers greater than zero were as common as dirt.*

So here's the main reason that Euler's formula is flabbergasting: the top five celebrity numbers of all time appear together in it with no other numbers. (In addition, it includes three primordial peers from arithmetic: +, =, and exponentiation.) This conjunction of important numbers, which sprang up in different contexts in math and thus would seem to be completely unrelated, is staggering, and it accounts for much of the hullabaloo about the equation.

Here's an analogy: Say that future astronomers identify scores of distant solar systems with planets that are almost exactly like Earth (right down to the level of oxygen in the atmosphere), and that in every such case the Earth-lookalike turns out to be the third planet out from its local star, and, furthermore, that the five planets nearest to the lookalike are nearly identical in all respects to Mercury, Venus, Mars, Jupiter, and Saturn—the five planets closest to Earth. This would be nothing short of astounding, and it would suggest the existence of a completely unanticipated, deep regularity in the structure of the universe. The tight-knit pattern of seemingly unrelated, supremely important numbers in Euler's formula is similarly provocative.†

---

*A version of zero had been used as a placeholder in number systems (a kind of spacer, that is, which is a role that zero plays in our decimal number system—for instance, to distinguish between 606 and 66) as far back as the ancient Babylonians. But it was later Indian mathematicians who are credited with making the conceptual leap of accepting nothing as something, number-wise. The fact that their Hindu religion included the concept of nothingness—the void—may account for the fact that this major conceptual leap took place in India.

† I feel obliged to mention here that some people feel that the equation isn't all that interesting. They have their reasons, which I disagree with and will address in the last chapter.

But that's not all that's surprising about it. Try another little thought experiment: Imagine that you'd never heard of Euler's formula but were familiar with the basics sketched above on *e*, *i*, and π. Now be honest—wouldn't you have expected $e^{i\pi}$ to be (a) gibberish along the lines of "elephant ink pie," or, if it were mathematically meaningful, to be (b) an infinitely complicated irrational imaginary number? Indeed, $e^{i\pi}$ is a transcendental number raised to an imaginary transcendental power.* And if (b) were the case, surely $e^{i\pi}$ would not compute no matter how much computer power were available to try to pin down its value.

As you know, neither (a) nor (b) is true, because $e^{i\pi} = -1$. (I suspect the fact that both (a) and (b) are provably false is the reason that Benjamin Peirce, the nineteenth-century mathematician, found Euler's formula (or a closely related formula) "absolutely paradoxical.") In other words, when the three enigmatic numbers are combined in this form, $e^{i\pi}$, they react together to carve out a wormhole that spirals through the infinite depths of number space to emerge smack dab in the heartland of integers. It's as if greenish-pink androids rocketing toward Alpha Centauri in 2370 had hit a spacetime anomaly and suddenly found themselves sitting in a burger joint in Topeka, Kansas, in 1956. Elvis, of course, was playing on the jukebox.

---

*If you were given to thinking of numbers as having human-like qualities, you might picture $e^{i\pi}$ as a guru into transcendental meditation who'd achieved infinite enlightenment. But there's a problem with that—Euler's formula shows that $e^{i\pi}$ can never free itself from worldly concerns. Recall, $e^{i\pi}$ is really −1 in disguise, and −1 is just a mathism for owing a dollar to your friend, Steve. One hand clapping.

CHAPTER 7

From Triangles to Seesaws

The key source of the wormhole-like surprise of Euler's formula is its imaginary-number exponent, *i* times π. Euler was the first to figure out how to interpret such strange exponents.

As we've seen, in the mid-1700s many mathematicians still regarded the imaginaries as iffy numbers—Euler himself said that they had an air of impossibility. Thus, imaginary exponents really pushed the envelope in math at the time. The very idea of raising a number to an imaginary power may well have seemed to most of the era's mathematicians like asking the ghost of a late amphibian to jump up on a harpsichord and play a minuet.

But Euler loved pushing envelopes. As math pioneers often do, he would manipulate well-accepted concepts and symbols to derive novel equations, then use the novelties to derive further math- and mind-expanding results. Using this strategy he showed that imaginary exponents could be translated into unexpectedly familiar terms.

Though wonderfully ingenious, the way he did that (which led to $e^{i\pi} + 1 = 0$) isn't hard to follow for those who have taken high school calculus. I'll give you a simplified version of it a bit further on, *sans* calculus.

But first a preview: Euler showed that e raised to an imaginary-number power can be turned into the sines and cosines of trigonometry.

Before you sue me for breach of contract ("It was fraud plain and simple, Your Honor—he promised easy math and the next thing I knew I was choking on trig."), let me offer you a no-choke mini-primer on sines and cosines.

Sines and cosines are functions. As noted earlier, functions resemble computer programs that take in numbers, manipulate them in some defined way, and output the results. But the trig functions are more interesting than simple functions such as $2x + 8$. They're like automated phonebooks that use people's names as input and output their phone numbers by looking them up in directories consisting of names paired with the phone numbers. (The conceptual roots of this analogy go back to, you guessed it, Leonhard Euler. He originated the fruitful idea of thinking about functions in general terms as coupled numbers. As he put it, functions are essentially quantities that "so depend on other quantities that if the latter are changed the former undergo change.")

The trig functions' input consists of the sizes of angles inside right triangles. Their output consists of the ratios of the lengths of the triangles' sides. Thus, they act as if they contained phone-directory-like groups of paired entries, one of which is an angle, and the other is a ratio of triangle-side lengths associated with the angle. That makes them very useful for figuring out the dimensions of triangles based on limited information. The word trigonometry comes from the ancient Greek for "triangle measurement."

To understand how the trig functions work, you need to know that a right triangle, as pictured in Figure 7.1, is one with a 90-degree angle and two smaller angles. The side opposite the right angle is called the hypotenuse. I've arbitrarily assigned the hypotenuse here a length of 1 unit. The units could be any measure of distance—millimeters, inches, miles, light-years, you name it.

FIGURE 7.1

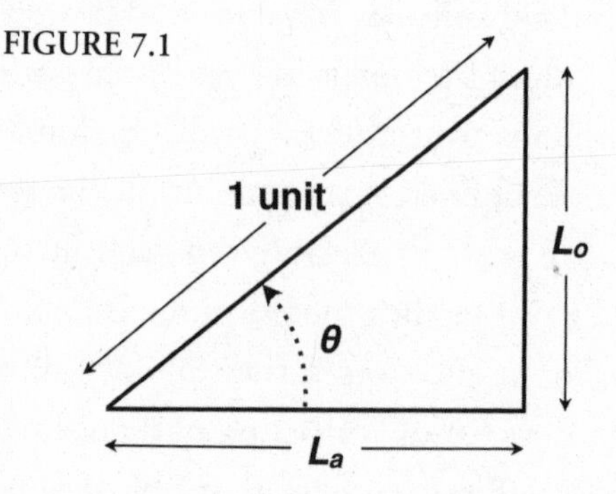

L_o stands for the length of the side that's opposite the angle marked θ, and L_a represents the length of the side adjacent to the angle θ. (The side that's not the hypotenuse, that is.)

The sine function is usually abbreviated as sin θ, where the Greek symbol θ (theta) is a variable representing the size of an angle. Note that θ has multiple roles here: it stands for the name of an angle as well as its size. Because it's a variable, you can plug in numbers for it in sin θ, and the function, in effect, will output other numbers in a precisely determined way. A typical input number would be the size, in degrees, of one of the two non-90-degree angles inside a right triangle. The sine function would then output *the ratio between the length of the side opposite to that angle and the length of the hypotenuse.*

For the triangle shown above, the sine of the angle θ, sin θ, equals the ratio of L_o to the length of the hypotenuse, which is 1. Expressed as a fraction, that's $L_o/1$, and since any fraction with a denominator of 1 can be reduced to its numerator alone (examples: 4/1 = 4 and 200/1 = 200), we have $\sin \theta = L_o/1 = L_o$. If we turn these stuck-together equations around and drop out the $L_o/1$ part, we get $L_o = \sin \theta$. This last equation means that the sine of the angle θ in the above triangle will tell us the length of the side marked L_o.

Now let's plug in a specific angle measurement for θ. The angle θ in the above triangle is about 38 degrees. My calculator says that the sine of that angle is approximately 0.616. (Calculators express fractions—which, recall, are conceptually equivalent to ratios—as the fractions' decimal equivalents.)

Therefore, we have $L_o = \sin 38° \approx 0.616$ units. (The squiggly equals sign in front of 0.616 means "approximately equal to.") Note that the sine function (that is, the version of it that exists in my calculator) let me determine L_o without the use of a ruler. This is an example of how trig functions can help size up a right triangle's dimensions based on limited information, which in this case was the size of its angle θ and the one-unit length of its hypotenuse.

The cosine function, written cos θ, is similar to the sine function, except that it outputs *the ratio between the length of an angle's adjacent side and the length of the hypotenuse.*

Thus, for the triangle shown above, cos θ = cos 38° = $L_a/1$ = L_a. Calling on my calculator again, I found out that cos 38° ≈ 0.788, and thereby determined that the length of the side adjacent to the angle θ is about 0.788 units.

Here's a simple exercise to help you get these trig definitions down. Draw a right triangle with a 1-foot-long hypotenuse on a standard 8½-by-11 sheet of typing paper. (You'll need to draw the hypotenuse near a diagonal between two of the sheet's corners in order to make it fit.) Use a protractor to measure one of the triangle's non-90-degree angles. (Don't have a protractor? Here's a work-around: if you draw the triangle's hypotenuse exactly on top of the sheet's diagonal, the angle between it and a triangle side drawn parallel to the sheet's shorter side will be about 52.3 degrees.) Then, following the procedure sketched above, use the trig functions to predict the lengths of the triangle's two sides (the ones other than the hypotenuse) in inches.*

To complete the exercise, measure the two triangle sides with a ruler to check whether your trig-based predictions are about right.

---

*Some helpful hints: If you don't have a calculator to look up the sine and cosine of an angle, you can use Google instead. Simply enter something like "sin 72 degrees =" (without quote marks) as the search term and the search engine will act like a calculator. If your triangle's hypotenuse is 1 foot long, the side lengths you calculate with the trig functions will be expressed in the same units, namely feet. To convert these lengths to inches, multiply them times 12. Then, when you measure the triangle sides with a ruler, convert fractions of an inch to their decimal equivalents so that both the trig-predicted lengths and the ones you measure with a ruler are expressed in decimals. To do the conversion you simply divide the denominator of each fraction into its numerator. (A fraction can be regarded as a little division problem in which the denominator is divided into the numerator to calculate the fraction's decimal equivalent.)

SINCE I'VE SPECIFIED plugging right triangles' angles into the sine and cosine functions, you might think that inputting angles greater than 90 degrees wouldn't be allowed. Indeed, if the angle θ in the above triangle were 90 degrees or more, there wouldn't even be a triangle to talk about.

But there's a clever way to expand the definitions of the sine and cosine functions to permit input angles greater than 90 degrees. As I'll explain, this change can be thought of as reprogramming the two functions so that their internal directories are based on angles swept out within a circle, rather than on angles within triangles.

To see how the expanded definitions work, you need to know a little about the *xy* plane, which sprang from the work of seventeenth century French philosopher and mathematician René Descartes. It's usually pictured as a flat, two-dimensional surface featuring a horizontal number line called the *x* axis and a vertical number line called the *y* axis.

Points on the plane, as shown in Figure 7.2, are specified by pairs of numbers inside parentheses, called *x* and *y* coordinates. The coordinates are used to locate points in much the same way that street intersections can be pinpointed on a map with numbered east-west streets and north-south avenues. For the coordinate pair (2,3), 2 is the *x* coordinate. It specifies how far from the *y* axis the point represented by (2,3) is and thus stands for a horizontal distance marched off along the *x* axis. The coordinate pair's second number, 3, is the *y* coordinate—it shows how far the point designated by (2,3) is from the *x* axis and so represents a distance marched off vertically along the *y* axis. The point where the axes meet is called the origin, and its coordinate pair is (0,0).

FIGURE 7.2

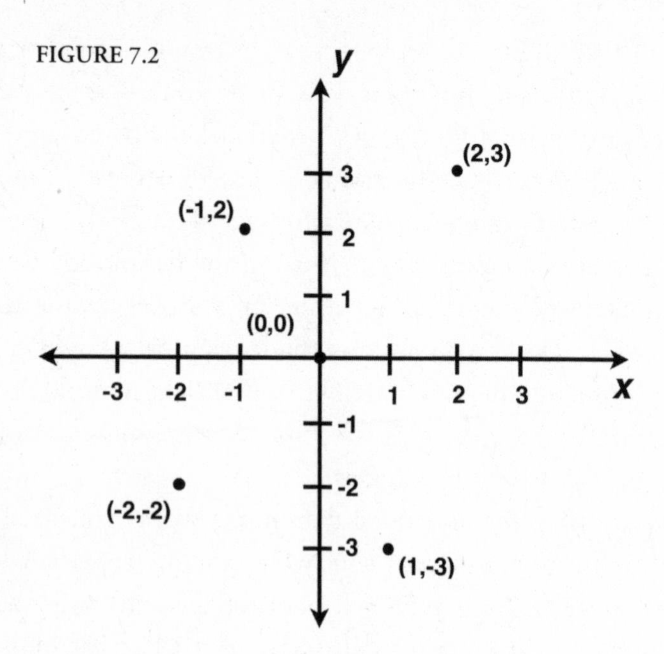

As you may remember from math classes, equations such as $y = 2x$ or $y = x^2$ can be represented as lines or curves on the xy plane by plotting points whose x and y coordinates satisfy the equations. This enables picturing functions geometrically, which often reveals things about them that would otherwise be hard to perceive. But that's analytic geometry, and we have trig fish to fry here, so let's get back to angles. Instead of pondering ones within triangles, however, we'll consider angles within a special circle at the center of the xy plane. It's called the unit circle.

(Historical aside: Trig was largely pioneered by astronomers who wanted to use it to determine distances to far-off objects such as the moon, and ancient Greek astronomer Hipparchus of Rhodes is credited with originating several of its

FIGURE 7.3

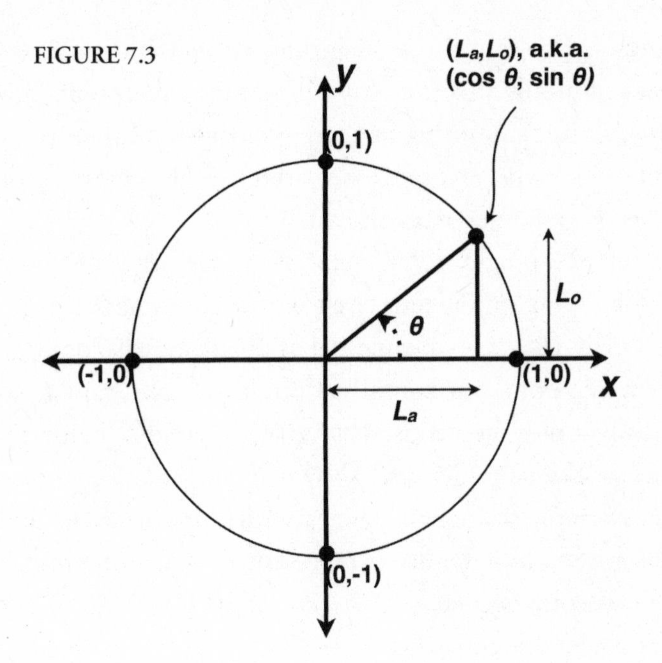

key concepts, including the idea that lengths of chords, lines drawn within a circle, are related to the lengths of arcs measured along its circumference—a forerunner to defining trig functions in terms of circles. But it was Euler who put the unit circle front and center in trig, and his doing so crystalized what mathematicians regard as the modern version of trigonometry. In the remainder of this chapter, I'll give you a simplified description of this version.)

The center of the unit circle is the point (0,0), the origin, and its radius is one unit in length. As shown in Figure 7.3, the unit circle intersects the x and y axes at the points (1,0), (0,1), (–1,0), and (0, –1).

A one-unit-long line segment akin to a clock's minute hand is usually drawn inside the circle to indicate the sweeping out of angles. Looking at the diagram, you should picture this segment initially positioned with its tip at the point (1,0)—that is, as if it's pointing at three o'clock—and then rotating counterclockwise around the origin. This rotary motion sweeps out an angle θ, as shown.

After sweeping out θ, the end of the segment is positioned at a point lying on the unit circle whose coordinate pair in the diagram is (L_a, L_o)—meaning that to locate the point, you measure L_a units horizontally along the *x* axis, and L_o units vertically along the *y* axis. You've seen L_a and L_o before: they represent the same lengths shown in the triangle above. In fact, that same triangle is delineated within the circle, and its one-unit-long hypotenuse is represented by the one-unit-long angle-sweeping segment. The triangle also includes the angle θ that was pictured earlier.

I've drawn the previously featured triangle within the circle to help show you how the triangle-based definitions of the trig functions can be recast in terms of the unit circle. The details on how that's done are explained in the next few paragraphs. Believe it or not, you're already an old hand at such mapping of math ideas from one context to another. For instance, when learning about angles, you mapped the idea of numbers onto the concept of angular distances between two intersecting lines—you probably didn't even notice yourself making this rather impressive conceptual leap.

The triangle's side along the *x* axis, which is L_a units in length, is adjacent to the angle θ of the triangle, just as it was in the earlier triangle diagram. Thus, $\cos\theta = L_a/1 = L_a$, based on the triangle-based definition of the cosine function. (This double equation should look familiar—it appeared above with the triangle.)

But L_a now has a second meaning, one that's related to the unit circle. That is, it designates the distance that must be marched off along the *x* axis to specify the *x* coordinate of the point on the circle at the tip of the angle-sweeping line segment. And because the two-part equation $\cos \theta = L_a/1 = L_a$ tells us that L_a is equal to $\cos \theta$, we know that this point's *x* coordinate can be written as $\cos \theta$ instead of L_a.

Similar logic applies to the *y* coordinate of the point at the tip of the angle-sweeping segment. That is, $\sin \theta = L_o/1 = L_o$, which means that $\sin \theta$ can be used as a proxy for L_o when writing the point's coordinate pair.

Therefore, as indicated in the diagram, the point's coordinate pair, (L_a, L_o), can also be expressed as $(\cos \theta, \sin \theta)$.

Now let's add some action to the scene. If the angle sweeper were initially positioned with its tip at the three o'clock position and then rotated counterclockwise, its tip could effectively single out any point on the unit circle between the three o'clock and twelve o'clock positions. Each such point would be associated with a certain angle, call it θ, between 0 and 90 degrees. And for each such point, you could draw a right triangle inside the circle containing an angle θ along with a one-unit-long hypotenuse identical to the angle sweeper. For instance, if θ were close to 90 degrees, the triangle you'd draw would be tall and skinny, with L_a approaching a length of 0 units, and L_o approaching one unit. And for that tall, skinny triangle with a hypotenuse that's one unit long, the definitions of the cosine and sine functions would guarantee that $\cos \theta = L_a$ and $\sin \theta = L_o$.

Conclusion: the coordinates of the point at the sweeper's tip can *always* be expressed as (cos θ, sin θ) when θ is between 0 and 90 degrees.

Notice that we have just opened up a new possibility for evaluating the cosine and sine functions for angles between 0 and 90 degrees: instead of outputting triangles' side-length ratios, the functions can output the x and y coordinates of points along the unit circle at the angle sweeper's tip after the sweeper designates an angle of θ degrees. In fact, if we redefined the functions so that their internal directories, so to speak, consisted of angles paired with these x and y coordinates (for cosines and sines, respectively), we wouldn't need to change the numbers in the functions' hypothetical directories at all. As we've just seen, they would contain the same input-to-output pairings regardless of whether they were based on right triangles or coordinate pairs along the unit circle.

BUT WHAT WOULD THE TRIG functions output if the angle sweeper got carried away and swept out an angle greater than 90 degrees?

This question can't be answered with reference to the kind of right triangle inside the unit circle that's shown above. The right triangle we'd need would have one right angle and another angle, call it θ again, larger than 90 degrees. There's no such triangle.

But fortunately our new circle-based alternative for specifying the functions' output can deal with this situation. That is, we can redefine the trig functions so that their output is based on the x and y coordinates of the point at the angle sweeper's tip after it sweeps out an angle greater than 90 degrees. (As well as less than 90 degrees.) And if we assume that the angle sweeper is moving in the negative direction when it rotates in a clockwise direction (like moving left of zero along the number line), the redefinition will even allow negative angles to be plugged into the sine and cosine functions. Who needs triangles?

Consider an example of the redefinition at work: Assume the angle sweeper has swept out a 180-degree angle, a half circle. Since by convention it always starts at the three o'clock position, it is now pointing in the nine o'clock direction—that is, at the point with coordinates (–1,0). (Take a look at the diagram above if you have trouble picturing this.) Therefore, under the new definitions of the trig functions, the cosine function's output when that angle is plugged in must be –1, and the output of the sine function for the angle must be 0. If we write that with equations, we have $\cos 180° = -1$ and $\sin 180° = 0$. This example, as you'll later see, is crucial to understanding Euler's formula.

Here's another mini-exercise: What are $\cos 90°$ and $\sin 90°$ equal to? How about $\cos 360°$ and $\sin 360°$?*

Recapitulating, we've entered a triangle-free trig zone in which the cosine and sine functions have been souped up to handle any angle as input. (This includes angles greater than 360 degrees, as I'll explain momentarily.)

Figure 7.4 shows the souped-up definitions applied to an angle between 90 and 180 degrees. Note that in this case $\cos \theta$ must be a negative number between 0 and –1. That's because

*Answers: $\cos 90° = 0$, $\sin 90° = 1$, $\cos 360° = 1$, $\sin 360° = 0$.

FIGURE 7.4

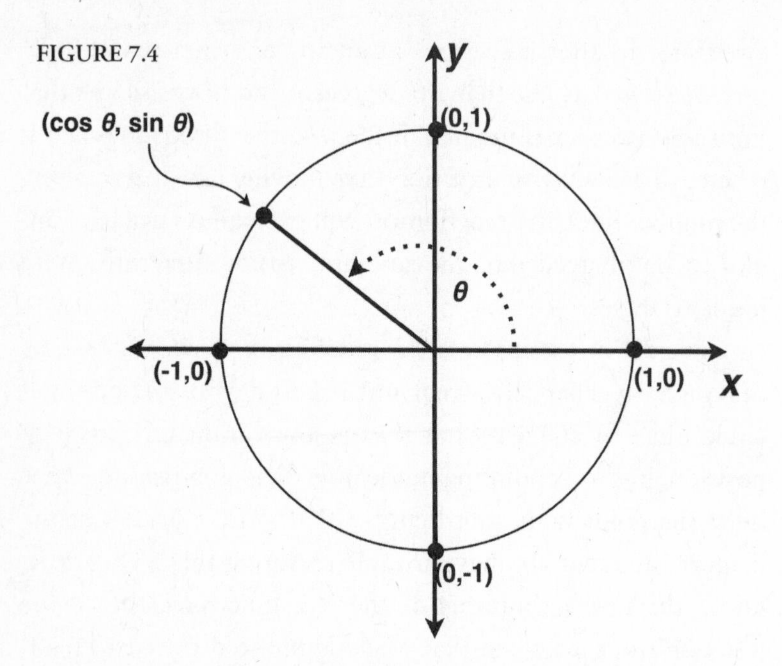

after the sweeper has swept out the angle, the x coordinate of the point at its tip is between 0 and –1, and, of course, cos θ is defined as the value of that coordinate. Similarly, sin θ for angles in that range is between 0 and 1, as I hope is apparent from the diagram. (If it's not, consider the tip point's y coordinate in light of where the point would be on the xy plane when the angle θ is in that range.)

Please try yet another exercise: determine the possible range of values for cos θ and sin θ when θ is greater than 180 degrees (a half circle) but less than 270 degrees (three-quarters of a circle), and when θ is greater than 540 degrees and less than 630 degrees.*

*Answer: In both cases, cos θ and sin θ will be between 0 and –1.

Hint on the second exercise with big angles: Sweeping out 360 degrees requires the angle sweeper to rotate all the way around the unit circle one time. And if it swept out, for example, 450 degrees, it would travel all the way around, back to where it started, plus 90 degrees beyond that point, because $450 = 360 + 90$. Thus, after sweeping out 450 degrees, the sweeper is positioned exactly where it would be if it had swept out only 90 degrees. It follows that the sine and cosine of 450 degrees are the same as the sine and cosine of 90 degrees. (That's because the trig functions' output for both angles is determined by the coordinates of the same point on the unit circle.)

Similarly, the sine and cosine of any angle will be the same as the sine and cosine of an angle that's 360 degrees less—or 2×360 degrees less, or 3×360 degrees less, or $n \times 360$ degrees less for any integer value of n. This implies that the sine and cosine of any angle θ greater than 360 degrees is equal to the sine and cosine of a readily visualized angle between 0 and 360 degrees, and you can calculate the latter angle by repeatedly subtracting 360 degrees from θ until the result is between 0 and 360 degrees. The fact that trig functions cyclically output the same numbers as ever larger angles are plugged into them, by the way, explains why endlessly undulating "sinusoidal" curves are produced when the sine and cosine functions are graphed in the *xy* plane.

If you had lots of time and a couple of measuring instruments (a protractor and a ruler), you could compile a table of the sine and cosine functions' outputs when various angles are plugged into them. (This table, by the way, could serve as a somewhat crude directory for looking up the outputs when angles are fed into the functions.) To put it together, you'd use the protractor to draw the angle sweeper inside a unit circle on the *xy* plane at precisely determined angles between 0 and 360 degrees (each of which would be swept out, as always, beginning at the three o'clock position). If you were very exacting, you might do this by one-degree increments around the entire circle. Then for each angle you'd determine the coordinates of the point at the segment's tip by carefully measuring the point's distances from the *x* and *y* axes. (You'd measure the distances along lines traced out between the points and the axes so that the lines intersected the axes at right angles.) Finally, you'd record the coordinates in your table as the values for the cosine and sine functions for that angle.

After you'd completed the table, you could use it to position the angle sweeper to indicate a particular angle—say 143 degrees—without using a protractor. To do that, you'd look up the values of cos 143° and sin 143° in your table in order to use them as the coordinates of the point that should be at the sweeper's tip after it sweeps out 143 degrees. Since cos 143° is equal to about −0.799, you'd measure 0.799 units in the negative direction (left from zero) on the *x* axis. And since sin 143° equals about 0.602, you'd measure that distance from the origin along the *y* axis. Based on these measurements, you'd make a dot at the point whose coordinates are (cos 143°, sin 143°), or (−0.799, 0.602). (It should be on the unit circle.) Finally, you'd draw the angle-sweeping line segment extending from the origin to the dot. The segment, as drawn, would reveal just how big the angle is.

One way to think about all this is to picture the coordinate pair ($\cos\theta$, $\sin\theta$) as a kind of device to automate the sweeping out of angles inside the unit circle. When you plug in an angle for θ, ($\cos\theta$, $\sin\theta$) makes the sweeper wind up exactly where it should be after rotating through θ degrees, by specifying the position of its tip point after it sweeps out that angle.

The redefined trig functions can be used to model **oscillation** (see box on next page), such as the cyclic movement of swings and seesaws (small kids are really into oscillation). The alternating current emanating from every plug in your home is also oscillatory, and electrical engineers have used gobs of trig since the late 1800s when designing AC-based circuits. By rights, a lot of the electrical devices in our lives should have little signs on the back that read "Trig Inside."

To put the finishing touch on this minimalist trig primer, I'll briefly introduce an alternative to degrees for measuring angles: radians. We need radians because angles plugged into the redefined sine and cosine functions (based on the unit circle) are usually measured in radians rather than in degrees.

Why radians? The reason is that specifying angles in radians often makes calculations easier than when degrees are used. The ancient Babylonians originated the use of degrees to measure angles—they liked the number 60 and its multiples (such as 6×60, or 360) so much that they based all their math on it. (It seems they were inspired by the fact that there are about 360 days in a year.) Thus, measuring angles with degrees, while often handy, is actually a vestige of an ancient way of thinking, and it entails needless clutter and complexity when used in higher math (and even in some lower math

Oscillation is broadly defined as back-and-forth motion. To see how oscillation is linked to trig functions, picture the angle sweeper inside the unit circle continually rotating like a clock's minute hand. With this image held in your mind, note that the sweeper is going back and forth between the three o'clock and nine o'clock positions as time passes. (Or between any two diametrically opposed points along the circle, for that matter.) This going back and forth is a form of oscillation, which means that oscillation can be represented in terms of rotary motion. Now imagine that the rotation is driven by plugging continually larger angles into θ in the coordinate pair ($\cos\theta$, $\sin\theta$)—the driver, so to speak, of the angle sweeper within the unit circle. If θ were increased by, say, 360 degrees per second, the point at the sweeper's tip would travel around the circle once a second. You could think of this trig-function-driven motion as oscillation at one cycle per second.

Historians credit Galileo with conceptually linking rotary motion to oscillation. But even before the great Italian scientist saw how the two are conceptually linked, anonymous sixteenth-century German inventors put the linkage to use in the first treadle-driven spinning wheels, which translate oscillatory up-and-down foot motion into rotary motion.

contexts)—degrees are like Roman numerals in this regard.* These hassles can be avoided by using radians.

---

*The militaristic, bullying Romans were so hopelessly inert at math that if the ancient world had had a no-civilization-left-behind program in mathematics they would have been stuck in the remedial category for well over a thousand years. As math historian Morris Kline put it, their "entire role in the history of mathematics was that of an agent of destruction." Most notably, they murdered Archimedes, the incandescent, irrepressible, eureka-shouting Feynman of the ancient world. At least there weren't nuke codes in those days for their Neros and Caligulas to get their hands on.

Radians fit hand-in-glove with the sweeping out of angles within the unit circle. When an angle sweeper moves along the circle to sweep out an angle, the point at its tip moves a certain distance along the circle. Radians are based on this arc length. Specifically, a radian is defined as the angle swept out by an arc along the circle that's equal in length to its radius. In the special case of the unit circle, the length of the radius is 1, and so the arc length associated with an angle of 1 radian is 1 unit.

How many radians would be swept out if the angle sweeper traveled all the way around the unit circle one time (that is, 360 degrees)? To get the answer, recall that the circumference of any circle is equal to its diameter times π. Since a circle's diameter is twice as long as its radius, the circumference equals $2 \times r \times \pi$ units, where r stands for the length of its radius, which is simply 1 for the unit circle. Therefore, the angle in terms of radians in this case would be $2 \times 1 \times \pi$, or, more compactly, 2π radians. Knowing that π is about 3.14, we could take this calculation one step further to conclude that the angle would be about 2×3.14, or 6.28, radians. But this step is very rarely taken in math—when radians are used to measure angles, they are almost always specified in terms of numbers multiplied times π.

Now that we've established that 360 degrees equals 2π radians, we know that half of 2π radians, or π radians, must be equal to half of 360 degrees, or 180 degrees. Similarly, half of π radians (180 degrees), or $\pi/2$ radians, must be equal to half of 180 degrees, or 90 degrees.

In sum, we have 2π radians = 360°, π radians = 180°, and $\pi/2$ radians = 90°. This is all you need to know about radians.

(If you have trouble remembering these radian/angle equivalences, just glance at the unit-circle illustration on the book's cover.)

As we saw earlier, cos 180° = –1 and sin 180° = 0. Since π radians = 180°, this implies that cos π = –1 and sin π = 0. (The word radians is typically omitted when writing angle sizes in trig functions.) These two trig facts will come into play when I show you how to bring forth Euler's formula.

Four other trig facts will also come up later. By examining the coordinates of the point at the sweeper's tip after it sweeps out π/2 radians (90°), you can see that cos π/2 = 0 and sin π/2 = 1. Lastly, cos 0 = 1 and sin 0 = 0, which follow from the fact if the sweeper moves zero radians, its tip remains at the point whose coordinates are (1,0).

IF YOU'VE BEEN ABLE to handle this chapter, you're ready to take possession of the once-hidden connection between trig functions and imaginary-number exponents that Euler uncovered. This connection quickly leads to Euler's formula and its fascinating implications. But before going into its details, in the next chapter I'll show you how trig functions, imaginary numbers, and infinity can be combined in a readily understood way to arrive at a pleasing result. It will introduce you to the kind of conceptual moves that Euler made as he explored the hidden trails leading to $e^{i\pi} + 1 = 0$.

CHAPTER 8

Reggie's Problem

Let's say that while you're at a party with friends, the conversation turns to what everyone has been reading lately. When you confess that you've been delving into math—imaginary numbers, trig, infinite sums, and the like—and to your surprise found it more understandable and interesting than you'd expected, the hosts' very bright but socially inept son, a 14-year-old math whiz named Reggie, is suddenly all ears. Excitedly butting into the conversation, he exclaims that he's been learning about exactly the same things and wonders if you'd like to try working an "easy" problem on them from his favorite book, *Basic Math for the Complete Genius*.

Before you can beg off, he whips out a piece of paper and scribbles the following scary-looking function consisting of an infinite sum:

$$f(\theta) = (i \cos \theta)^2/2 + (i \cos \theta)^4/4 + (i \cos \theta)^8/8 + (i \cos \theta)^{16}/16 + \cdots.$$

"Here's the problem," he says. "What's this function's value when the variable equals pi?

"It took me two minutes to figure it out," he adds brightly, oblivious to the blush spreading across your face. "It's really simple. Even Jake—he's my math club friend who isn't as good as me at solving problems—figured it out pretty fast."

Thinking to yourself, "What an annoying little dweeb," you say, "Oh, I'm sure I'm not as good at math as you and Jake. But I'll take a crack at it later. Thanks for sharing it with me."

Picking up the piece of paper before tucking it into your pocket, you make a little show of perusing what Reggie has written so as not to seem dismissive of your hosts' son. Then suddenly you have a flash: when you substitute π for θ in the infinite sum, every $\cos \theta$ will become $\cos \pi$, which is equal to -1. (In fact, while reading this book's trig chapter recently, you learned that $\cos \pi = -1$.) That means all the cosines will go away, leaving behind unscary -1's. In addition, you recall that i^2 is equal to -1 (since i is defined as the square root of -1), and so whenever "i times i" appears in any of the fractions, it can also be replaced by -1.

So you think, "Maybe this isn't as bad as it looks." Instead of pocketing the problem, you find yourself asking Reggie to lend you a pencil so you can sit down to fiddle with it. Everyone is astonished, including you. But then, you never did like doing what people expected you to do—beneath your facade of polite good humor lurks a wild thing. (Why else would you be reading this book?)

You start by evaluating the infinite sum's first θ-containing term, $(i \cos \theta)^2/2$, when θ is set equal to π. The $i \cos \theta$ part represents i multiplied times cos θ. Thus, after plugging π in for θ, you write, using ×'s for multiplication,

$(i \cos \pi)^2 = (i \times \cos \pi) \times (i \times \cos \pi)$ [by the definition of exponents]
$= (i \times -1) \times (i \times -1)$ [by subbing in -1 for $\cos \pi$]
$= i \times i \times -1 \times -1$ [rearranging by the commutative and associative laws]
$= i^2 \times 1$ [definition of exponents, and enemy of my enemy = friend]
$= -1 \times 1$ [i is defined as the square root of -1, and so $i^2 = -1$]
$= -1$ [the enemy of my friend is someone I feel negative about].

You're on your way: Since you've shown that the numerator of the fraction $(i \cos \pi)^2/2$ is equal to -1, you've proved that the fraction is just an absurdly complicated way of writing $-1/2$.

Now you fearlessly attack the numerator of the second term, $(i \cos \pi)^4/4$:

$(i \cos \pi)^4 = (i \times \cos \pi) \times (i \times \cos \pi) \times (i \times \cos \pi) \times (i \times \cos \pi)$
$= (i \times \cos \pi)^2 \times (i \times \cos \pi)^2$ [definition of exponents]
$= -1 \times -1$ [since, as shown above, $(i \times \cos \pi)^2 = -1$]
$= 1.$

Which means that $(i \cos \pi)^4/4$ equals $1/4$.

After a moment's thought, you realize that the numerator of every subsequent fraction is of the form $(i \times \cos \pi)$, with an even exponent larger than 4, meaning that it can be reduced to an even number of terms of the form $(i \times \cos \pi)^2$ multiplied together—this follows from the same logic used above to simplify $(i \cos \pi)^4$. Those terms, in turn, can be reduced to pairs of –1's multiplied together, and since each such pair is equal to 1, what you've got in each numerator is just 1's multiplied together, or simply 1. For instance,

$$\begin{aligned}(i \cos \pi)^8 &= (i \times \cos \pi)^2 \times (i \times \cos \pi)^2 \times (i \times \cos \pi)^2 \times (i \times \cos \pi)^2 \\ &= -1 \times -1 \times -1 \times -1 \\ &= 1 \times 1 \\ &= 1.\end{aligned}$$

Thus, $(i \cos \pi)^8/8 = 1/8$.

And so on.

Putting it all together, you write Reggie's original equation with π plugged in for θ,

$$f(\pi) = (i \cos \pi)^2/2 + (i \cos \pi)^4/4 + (i \cos \pi)^8/8 + (i \cos \pi)^{16}/16 + \cdots$$

and then below that write the partial solution to the problem that you've now figured out:

$$f(\pi) = -1/2 + 1/4 + 1/8 + 1/16 + \cdots.$$

Looking on, Reggie says in his piercing way, "That's not the answer. You have to figure out what the infinite series is equal to."

"Thanks for pointing that out," you reply, miraculously sounding as if you mean it. By this time, your hosts and other friends are crowded around to see what you're doing. Obviously you can't stop now.

Fortunately, you're on a roll: You recall running across an infinite sum of fractions that looked a lot like the one you're now facing. Let's see, when was that? Then it comes to you—it was the time you read about Zeno's paradox.*

How did it go? Oh yes—an ancient Greek philosopher named Zeno pictured a runner on a racetrack. First, the runner went half of the way to the finish line, then half of the remaining distance (equal to a fourth of the way), then half of the remainder after that (an eighth of the way), and so on. It appeared that he couldn't possibly get to the finish given the infinite number of segments he had to complete.

But it seemed there was a flaw in Zeno's argument. The Greek philosopher implied that it should take an infinite amount of time for the runner to complete the race because he had to cross an infinite number of racetrack segments, each of which would have required some amount of time to negotiate. But Zeno apparently wasn't aware of the fact that an infinite sum of fractions can be finite if the successive fractions dwindle toward zero. In fact, the sum of racetrack segments that the runner had to cross (1/2 + 1/4 + 1/8 + ···) equals 1, not infinity. Your source on Zeno had shown you

*Zeno, the infinity-invoking puzzle poser mentioned in Chapter 3, came up with several related paradoxes that have stirred debate for about 2,500 years. Some mathematicians and philosophers have argued that the paradoxes are based on naïve or faulty premises and thus aren't truly paradoxical. But others have argued that purported solutions to the paradoxes have merely glossed over deep issues they raise. In any case, the lack of universal agreement about how to deal with Zeno's conundrums suggests that their power to perplex still hasn't been definitively quashed.

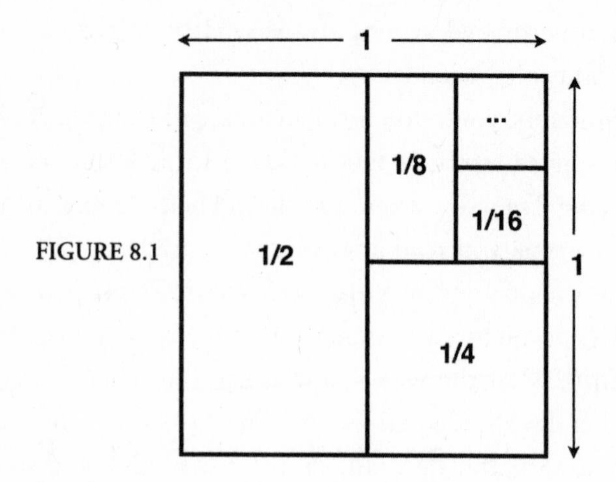

FIGURE 8.1

how to demonstrate that by drawing a square, 1 unit on a side, divided into subparts—as shown in Figure 8.1.

The total area of this square is 1, since the area of any rectangle equals length times width, and both the length and width are both 1 in this case. Now, when half of its area is added to half of the remaining area, which is one-fourth of the square, and then half of the area remaining after that (which is one-eighth) is added to the two previously summed areas, and so on, the sum of areas will get ever closer to the total area of the square. In fact, the difference between the sum and the total area can be made ever smaller by adding enough fractional areas to the sum. In other words, the difference can be reduced to a reasonable facsimile of zero, which means that the infinite sum and the total area are actually equal.

This implies that one-half plus one-fourth plus one-eighth and so on forever add up to one. (And that means, by the way, that if the runner goes at a constant speed of one racetrack length per x minutes, it will take him exactly x minutes to cross the infinite sum of fractions to get to the finish, since they add up to one racetrack length.)

Now nothing can stop you. You recreate the square on the piece of paper for everyone to see. Then you triumphantly exclaim, "The areas of all the subparts of this square add up to the total area of the square, which we know is 1. This shows that

$$1/2 + 1/4 + 1/8 + \cdots = 1.$$

"If we add -1 to both sides of this last equation, we get

$$-1 + 1/2 + 1/4 + 1/8 + \cdots = -1 + 1$$

and since the $-1 + 1/2$ on equation's left side is equal to $-1/2$, and the $-1 + 1$ on its right side is equal to 0, we can rewrite the equation as

$$-1/2 + 1/4 + 1/8 + \cdots = 0.$$

"Remember that we earlier found out that $f(\pi)$ is equal to an infinite sum of fractions:

$$f(\pi) = -1/2 + 1/4 + 1/8 + \cdots.$$

"Notice that this infinite sum is the same infinite sum that we just proved is equal to 0. In other words, $f(\pi)$ equals an infinite sum that is equal to zero, and so $f(\pi)$ itself is equal to zero. Thus, the answer to Reggie's problem is

$$f(\pi) = 0.$$

"There you go, Reggie. QED and all that. You were right, it isn't very hard."

CHAPTER 9

Putting It Together

Euler's discovery of a surprising connection between trigonometry and imaginary-number exponents wasn't the first example of a link between trig and the imaginaries. In the early 1700s, French mathematician Abraham de Moivre effectively constructed a bridge between these two math topics by originating a variant of this equation, now known as de Moivre's formula (although he didn't write it this way):

$$(\cos \theta + i \sin \theta)^n = \cos (n\theta) + i \sin (n\theta)$$

where n is an integer, θ stands for an angle measured in radians, $n\theta$ means n times the variable θ, and $i \sin \theta$ means i times $\sin \theta$.

At first glance, de Moivre's formula may look like a baffler from Reggie's favorite book. But it's not hard to understand. Just think of each side of the equation as a function that requires a two-step process to deliver an output number. First, you set n equal to a whole number on both sides. Then, you plug in a certain number of radians for θ in the two functions. The equation implies that the function on the left side will output the same number that the function on the right side does for that θ.

To see how it works, let's plug in 2 for n, and an angle of $\pi/2$ radians for θ, and then evaluate each side of the equation to see if they're really equal, as claimed by the formula. (If they're not, an awful lot of math books will need to be corrected.)

First, the left side:

$(\cos \pi/2 + i \sin \pi/2)^2 = (0 + (i \times 1))^2$ [since $\cos \pi/2 = 0$ and $\sin \pi/2 = 1$]
$= i^2$
$= -1.$

Now the right side:

$\cos (2 \times \pi/2) + i \sin (2 \times \pi/2) = \cos \pi + i \sin \pi$
$= -1 + (i \times 0)$
$= -1.$

So it seems, based on very limited evidence, that de Moivre got it right. (Appendix 1 offers stronger evidence—a derivation of the formula.) De Moivre is credited with a number of other advances besides his eponymous formula. While consulting with gamblers, for example, he developed important concepts in probability. Unfortunately, he never found a way to make a decent living. Despite efforts by celebrated contemporaries such as Leibniz to get him a university job, he was forced to scrape by as a private math tutor and spent his life in poverty. He was a refugee too—reared in France as a Protestant, he moved to London at age 20 to escape the persecution of Protestants during the reign of Louis XIV and lived the rest of his life in England.

In his old age, as the story goes, he noticed that he was sleeping 15 minutes longer every night and predicted that he would die on the day that his progressively lengthening period of sleep reached 24 hours. Being mathematically adept, it wasn't hard for him to calculate exactly when that would happen. Once again, and for the last time, he was right.

De Moivre's formula is a versatile tool in mathematics. But its full potential to light the way forward on a number of questions wasn't apparent until Euler, with his genius for making connections, got interested in it in the mid-1700s—apparently he derived it independently of de Moivre. One of Euler's intriguing findings based on the formula was that the sine and cosine functions are akin to jack-in-the-boxes with infinity coiled up inside. That is, he showed that the trig functions are equal to functions consisting of infinite sums. These sums, made up of fractions of θs raised to successively greater powers, possess orderly patterns of beautiful simplicity based on even and odd integers. Take a look:

$$\cos\theta = 1 - \theta^2/2! + \theta^4/4! - \theta^6/6! + \theta^8/8! + \cdots$$

and

$$\sin\theta = \theta - \theta^3/3! + \theta^5/5! - \theta^7/7! + \theta^9/9! + \cdots.$$

The ! symbols in these two equations stand for **factorials.** The factorial operator has explosive power, quantitatively speaking—it can expand even smallish numbers into astronomically huge ones. For instance, 15! is over 1.3 trillion. That means the successive denominators of the fractions in these two equations get bigger very fast, which in turn means that the successive fractions themselves become vanishingly small very fast.

The factorial symbol, !, is shorthand for "multiply together all the positive integers up to and including the specified integer." Thus, 3!, which is spoken "three factorial," is short for 1 × 2 × 3, or 6. And four factorial, or 4!, means 1 × 2 × 3 × 4, which equals 24.

Interestingly, de Moivre's formula includes imaginary-number terms—the ones with *i* times sines—while the infinite sums Euler derived from the formula feature only real numbers. Thus, he effectively traveled through the land of the imaginaries to get to new results in the real-number realm. During the course of his derivation, the imaginary-number terms that he began with in de Moivre's formula disappeared, just as all the *i*'s did in the evaluation of Reggie's infinite sum. The ability of imaginaries to suddenly disappear in calculations, as when i^2 becomes −1, is probably why Leibniz regarded them as slippery little creatures halfway between being and not-being.

These equations make it possible to determine values for the sines and cosines of particular angles without measuring triangles' side lengths or going through the laborious protractor-and-ruler process with the unit circle outlined in the trig chapter. To calculate the approximate cosine of a particular angle, for example, you could simply plug in the angle for θ in the first few terms of the infinite sum of the first equation and add them together. (You need to use only the first few fractions of the infinite sums because, as noted above, the factorials in the denominators cause the successive fractions to dwindle very rapidly toward vanishingly small amounts—only the first few are large enough to have a significant effect on the total.)

Let's try that for an angle of 38°, whose approximate cosine was obtained via calculator in the trig chapter: cos 38° ≈ 0.788. Of course, performing this exercise won't prove that the infinite sum Euler found hidden inside the cosine function is valid. But let's hope that it lends a bit of circumstantial evidence in favor of that conclusion.

Euler's equations refer to the unit-circle-based sine and cosine functions, and so we must convert 38° to radians before plugging it in for θ. Since we know that 180 degrees equals π radians, 38° should be equal to 38/180 as many radians as that. Because π equals about 3.14, we have 38° ≈ 38/180 × 3.14 radians, which means 38° is about 0.663 radians.

Plugging this radian-expressed angle into the first of the two equations gives

$$\cos 0.663 = 1 - (0.663)^2/2! + (0.663)^4/4! - (0.663)^6/6! + (0.663)^8/8! + \cdots.$$

Adding up the first five terms in the infinite sum yields cos 38° ≈ cos 0.663 radians ≈ 0.788. Happily, my calculator's estimate of cos 38° agrees with this.

Plugging in 0.663 for the angle in the sine equation's first five terms gives

$$\sin 0.663 = 0.663 - (0.663)^3/3! + (0.663)^5/5! - (0.663)^7/7! + (0.663)^9/9!.$$

Adding up the terms on the right side of the equals sign yields sin 38° ≈ sin 0.663 radians ≈ 0.616—further reassuring agreement with my calculator.

OUR MAIN ATTRACTION is now just around the bend—recasting the trig functions as functions made up of infinite sums is a key step in the derivation of the most elegant equation. But we need one last venture into the infinite to reach it, a jaunt revealing that the function e^x is another jack-in-the-box with a beautifully patterned infinite sum inside.

Before showing what popped out of e^x when Euler turned its crank, I should point out that the derivation of Euler's formula that I'm sketching in this chapter is just one of three ways that Euler demonstrated the truth of a general equation from which $e^{i\pi} + 1 = 0$ follows as a special case. The general equation, which confusingly is also often called Euler's formula, is

$$e^{i\theta} = \cos \theta + i \sin \theta.$$

(Many math books state this formula using x instead of θ as the variable, by the way.) I'll explain where it comes from in the next few paragraphs and further expand on its meaning in the following chapters. The derivation that I'm discussing here is the one Euler came up with that most closely resembles those typically shown in math texts today. Another of his derivations, however, is arguably the easiest to follow for those who aren't familiar with calculus although it's unorthodox by modern standards; that entire derivation is shown in Appendix 1.

Here's the infinite sum that Euler found inside e^x, which, to be consistent with this chapter's use of θ as a variable, I'll write as e^θ:

$$e^\theta = 1 + \theta + \theta^2/2! + \theta^3/3! + \theta^4/4! + \theta^5/5! + \cdots.$$

This infinite sum may look at least somewhat familiar. In fact, if you edited the two trig functions' infinite sums shown above by replacing their minus signs with plus signs, and then added the two infinite sums together, you'd get precisely the same infinite sum that Euler popped out of e^θ. This equivalence of infinite sums foreshadows his revelation that *e* raised to an imaginary-number power can be expressed in terms of sines and cosines.

To bring imaginary exponents into the picture, Euler made what modern mathematicians regard as a very bold move: He replaced all the θs in the above equation for e^θ with an imaginary-number version of θ. This step is now viewed as a kind of Evel Knievel leap because after Euler had proved that the equation was true for real-number θs, he basically just assumed that it would also be true when imaginary numbers were plugged in for the θs. Although he didn't base this leap on a rigorous argument, it was a sound assumption—Euler's intuitive moves were usually correct.

The imaginary-number version of θ that he subbed in is written $i\theta$, which means i times the variable θ —$i\theta$ is just the imaginary-number counterpart of the real-number variable θ. The substitution entailed rewriting e^{θ} as $e^{i\theta}$, as well as replacing all the θ's in the infinite sum on the right side of the equation with $i\theta$.

After these substitutions, the equation becomes

$$e^{i\theta} = 1 + i\theta + (i\theta)^2/2! + (i\theta)^3/3! + (i\theta)^4/4! + (i\theta)^5/5! + \cdots.$$

Note that the infinite sum on the right side is reminiscent of the one in Reggie's problem. Indeed, simplifying it by rewriting its exponential terms is actually easier than solving Reggie's problem. Let's do that.

We'll begin, once again, with the fact that i^2 is equal to -1. That means we can substitute -1 for each occurrence of i^2 in the numerators of the infinite sum's fractions. For instance, $(i\theta)^2$—the third term's numerator—equals $i\theta \times i\theta$, which, by rearranging the i's and θ's, equals $i^2 \times \theta^2$, or $-1 \times \theta^2$, which can be written $-\theta^2$. Thus, the third term is equal to $-\theta^2/2!$.

Similarly, the fourth term's numerator, $(i\theta)^3$, is equal to $i^3 \times \theta^3$, or $i^2 \times i \times \theta^3$ (since i^3 equals $i^2 \times i$), which equals $-1 \times i \times \theta^3$, which, by omitting multiplication signs, is equal to $-i\theta^3$. Thus, the fourth term can be simplified to $-i\theta^3/3!$.

Now that you've seen how this process works, try it on the next six terms of the infinite sum, similarly substituting -1 for all occurrences of i^2, and also replacing each occurrence of -1×-1 with 1. What do you get?*

---

* Answer: $\theta^4/4!$, $i\theta^5/5!$, $-\theta^6/6!$, $-i\theta^7/7$, $\theta^8/8!$, $i\theta^9/9!$.

These simplifying moves show that the above equation can be rewritten as

$$e^{i\theta} = 1 + i\theta - \theta^2/2! - i\theta^3/3! + \theta^4/4! + i\theta^5/5! - \theta^6/6! - i\theta^7/7! + \theta^8/8! + \cdots.$$

Note that every other element of the infinite sum is now a plus or minus term of the form i times $\theta^n/n!$, where n is an odd integer. (The first such term, $i\theta$, equals i times $\theta^1/1!$, because by definition a number raised to the first power is just that number, unchanged. In addition, 1! is defined as 1. Thus, $i\theta = i\theta^n/n!$ when $n = 1$.) Meanwhile the other terms are similar fractions with no i's in them. Let's rearrange the sum so that these two different kinds of terms are grouped together:

$$e^{i\theta} = [1 - \theta^2/2! + \theta^4/4! - \theta^6/6! + \theta^8/8! + \cdots] + [i\theta - i\theta^3/3! + i\theta^5/5! - i\theta^7/7! + \cdots].$$

Finally, we apply an expanded version of the distributive law* to the second group of terms so that each of their i's is accounted for by a single i multiplied times all of the group's terms. That justifies the following rewrite of the equation:

---

*The distributive law, a basic rule of arithmetic, is usually written $a \times (b + c) = (a \times b) + (a \times c)$, where the letters stand for numbers or more complicated expressions representing numbers. It means that when you multiply a sum by a number, you get the same result that you would get if you separately multiplied the number times each of the summed numbers and then added the products together. It can also be written as $(a \times b) + (a \times c) = a \times (b + c)$ by reversing the two sides of its standard formulation. And it can be extended to work with any number of terms. For instance, $(2 \times 4) + (2 \times 2) + (2 \times 7) + (2 \times 3) = 2 \times (4 + 2 + 7 + 3)$. Here, Euler adventurously applied it with an infinite number of terms.

$e^{i\theta} = [1 - \theta^2/2! + \theta^4/4! - \theta^6/6! + \theta^8/8! + \cdots] + [i \times (\theta - \theta^3/3! + \theta^5/5! - \theta^7/7! + \cdots)]$.

To get to the general version of Euler's formula, all we need to do now is to notice that the first infinite sum, in brackets, is equal to the infinite sum for cos θ that was shown earlier, and the second infinite sum, in parentheses, is equal to the infinite sum for sin θ. Thus, we can replace the infinite sums on the right side of the equation with the trig functions that they're equivalent to in order to arrive at the equation we're after,

$e^{i\theta} = \cos\theta + i\sin\theta$.

To recapitulate, the equation $e^{i\theta} = \cos\theta + i\sin\theta$ follows from the fact that the infinite sum shown above for $e^{i\theta}$ is equal to the infinite sum for cos θ plus i times the infinite sum for sin θ. As we've seen, Euler arrived at this remarkable equation by following a kind of concealed trail through the realm of the infinite to show that the equation's two sides, which seem radically different functions at first glance, are actually identical. (Meaning that when you plug in a number for θ in $e^{i\theta}$, it outputs the same number that $\cos\theta + i\sin\theta$ does when that θ is plugged in.)

GETTING TO MATHEMATICS' most beautiful equation from here is now a stroll in the park. First, plug in π for all the θ's in $e^{i\theta} = \cos\theta + i\sin\theta$. The substitution yields

$e^{i\pi} = \cos\pi + i\sin\pi$.

Because $\cos \pi = -1$ and $\sin \pi = 0$, you can substitute –1 for $\cos \pi$, and 0 for $\sin \pi$ to get

$$e^{i\pi} = -1.$$

(The $i \sin \theta$ term disappeared because after π is substituted for its θ, it becomes i times $\sin \pi$, which equals i times 0, and zero times any number equals zero.)

Then, by adding 1 to both sides of $e^{i\pi} = -1$ you get another equation in which both sides are equal:

$$e^{i\pi} + 1 = -1 + 1.$$

And, as the opening crescendo of "Also Sprach Zarathustra"* suddenly wells up out of nowhere, you simplify this equation to

$$e^{i\pi} + 1 = 0.$$

*You know, that stirring music from the movie *2001: A Space Odyssey*.

CHAPTER 10

A New Spin on Euler's Formula

People often invest new meanings in great works of art over time, breathing new life into them. The same thing happens in mathematics. A few years after Euler died, an obscure Norwegian surveyor with a gift for math conceived an innovative way of thinking about imaginary numbers that threw new light on Euler's formula. It's called the geometric interpretation of complex numbers.

The Norwegian innovator was Caspar Wessel, an amateur mathematician who always had trouble making ends meet with his wretchedly low-paid surveying work. An acquaintance described him in a letter as having "a bright, but very slow head, and when he sets out to study something, he can have no peace before he completely understands it." Wessel also struck those who knew him as distinctly meek: "If he had been in possession of more courage and assurance when it comes to trying unaccustomed work, then with his insight and talent, he could have done a lot for the benefit of the community as well as for himself," a fellow surveyor once wrote of him.

Wessel might never have published his important new ideas if mathematician Johannes Nikolaus Tetens, a prominent member of the Royal Danish Academy of Sciences, hadn't encouraged him. In 1797, Tetens, acting on behalf of his shy protégé, read a treatise by Wessel at a meeting of the academy. It was the only math paper Wessel wrote, and it laid out the geometric interpretation.

Unfortunately, Tetens failed to follow up by drawing wider attention to Wessel's paper, which was modestly, and somewhat mystifyingly, titled (when translated into English) "On the Analytical Representation of Direction." As a result it languished in obscurity for nearly a century despite its publication in the Danish academy's journal in 1799. Finally, in 1895, translations of it became widely available, showing that Wessel, who died in 1818, deserved credit for first developing the geometric interpretation.

A few years after Wessel's big idea appeared in the Danish journal, another amateur mathematician, Jean-Robert Argand, a Paris accountant, independently came up with the same advance. In a remarkable coincidence, his brainchild was also nearly lost to history. For some reason Argand decided to tell the world about it in an anonymous essay he self-published in 1806. It apparently made little impression on the few who saw it when it first came out. Luckily, seven years later French mathematician Jacques Français became aware of the work and was struck by its originality. Using a bit of detective work, he managed to get in touch with its

FIGURE 10.1

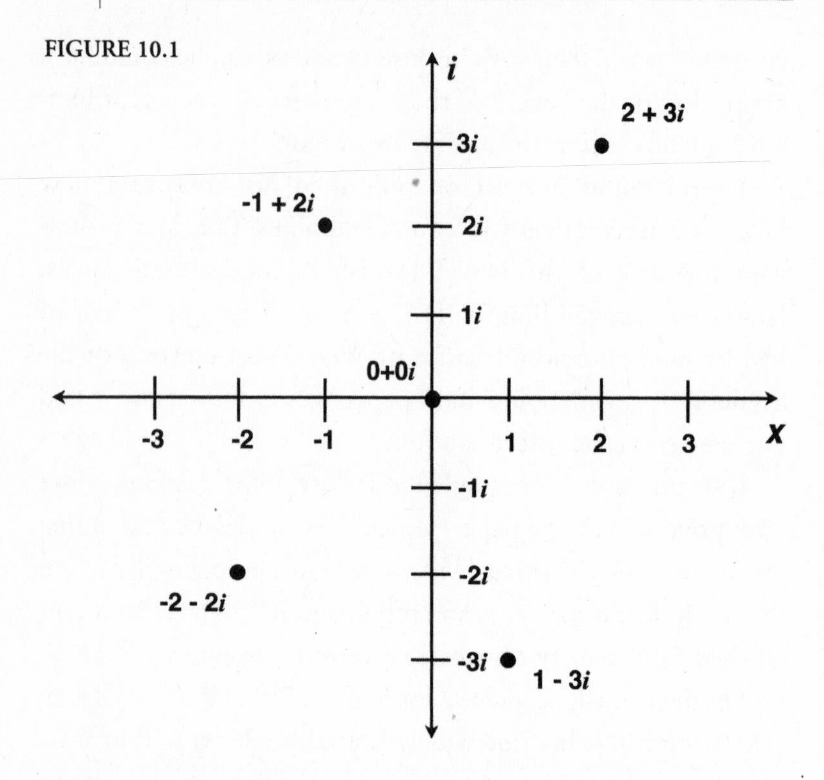

author and alerted the world to his innovation—Argand's name has been prominently associated with the geometric interpretation ever since.

In the geometric interpretation, the imaginaries are assigned to their own number line, called the *i* axis, which is drawn vertically, along with a real number line, called the *x* axis, which is drawn horizontally intersecting the *i* axis. As shown in Figure 10.1, the two intersecting number lines inhabit a flat space that looks very much like the *xy* plane, except that the *y* axis is replaced by the *i* axis.

The diagram shows why it makes sense to say the imaginaries hail from a different dimension: the *i* axis forms one of the two dimensions of a 2-D plane; the plane's other dimension is formed by the familiar real-number line, which is called the x axis in this context. This 2-D space is known as the complex plane, and the points on it are associated with two-part real-plus-imaginary numbers called complex numbers. Just as grade-school math's number line suggests how to think of the real numbers as points in 1-D space (lines have one dimension), the complex plane enables complex numbers to be mapped onto points in 2-D space.

Recall that points in the xy plane are designated with pairs of numbers called coordinates. Two numbers also are used to specify each point on the complex plane—a real one marched off horizontally along the x axis, and an imaginary one marched off vertically along the i axis. They're usually written, as shown in the diagram, in the form of sums, such as $2 + 3i$ or $1 + -3i$ (the latter complex number is the same as $1 - 3i$). Gauss, the German mathematician, is credited with coming up with this standard $a + bi$ format for writing complex numbers in the early nineteenth century, although some historians trace it back to Cardano, the sixteenth-century Italian mathematician who regarded doing arithmetic with imaginary numbers as mental torture.

Note that any real number can be viewed as a complex number whose imaginary part is 0 times i. The real number 2, for example, is essentially just a stripped-down version of $2 + 0i$. Since pure real complex numbers such as 2, or $2 + 0i$, have zero imaginary components, their points all lie on the real-number x axis in the complex plane. This makes sense because complex numbers' imaginary components specify how far away from the x axis their points are in the plane. That distance is zero when the imaginary parts are $0i$, and so the points for pure real complex numbers are zero distance away from the x axis.

Similarly, imaginary numbers such as $2i$ can be viewed as complex numbers with real-number parts equal to 0. The points for all such pure imaginary complex numbers lie along the i axis in the complex plane.

COMPLEX NUMBERS REPRESENT a major upgrade of the number concept. A complex number such as $2 + 3i$ incorporates two quantities, and thus it effectively holds twice as much information as a real number such as 2, or an imaginary number such as $3i$. That means complex numbers can do things that one-part numbers can't. You've already seen one example of this: while a number like 2 can be used to locate a point along a line, a complex number like $2 + 3i$ can be used to locate a point in 2-D space. Thus, the upgrade expands the concept of numbers from the realm of yardsticks to the richer world of maps. (Which, not coincidentally, is the world in which surveyor Caspar Wessel spent his career.) A real number could tell us how far we've come down a road, while a complex number could tell us exactly where we are.

The two-ness of complex numbers also enables them to simultaneously represent moving objects' speed and direction of motion. To see how this works, picture a field several acres in size. Now envision an archer at the field's center, and imagine that she shoots an arrow in a northeasterly direction at a certain speed—say 200 feet per second. To model the arrow's (bow-release) speed and direction with a single complex number, you could think of her as standing at the origin of the complex plane shown in Figure 10.2, with the *i* axis running north and south, and the *x* axis running east and west. By drawing an arrow on the plane extending from the origin in the direction of the point 1 + *i*, you could represent the real arrow's direction, northeast. And if you set the arrow's length to 200 units, you could represent the real arrow's speed by means of that length.

FIGURE 10.2

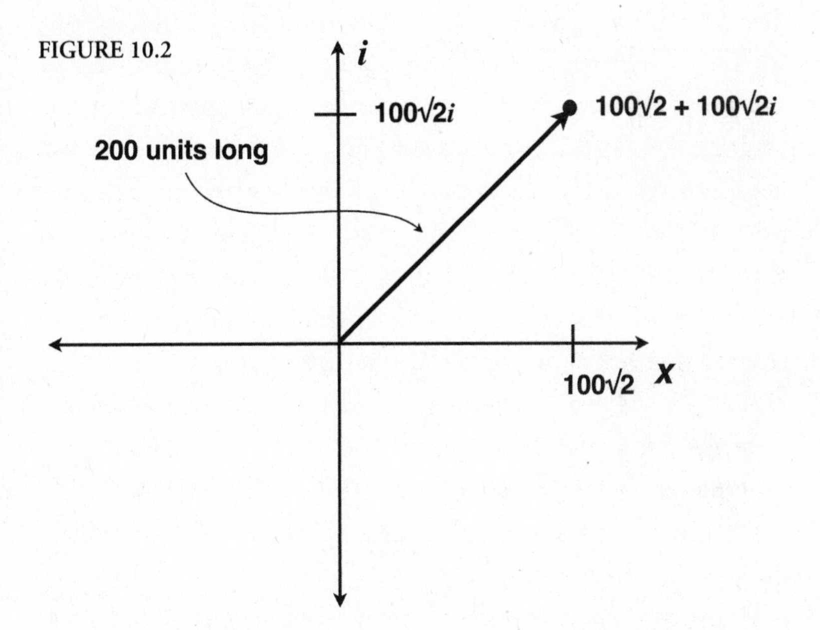

All the information contained in the arrow that you've drawn (the speed and direction of the real-world arrow) is encapsulated by the complex number for the point at its tip, which is $100\sqrt{2} + 100\sqrt{2}i$. ($100\sqrt{2}$ means 100 times the square root of 2.) In fact, if you were given that complex number and asked to figure out the speed and direction of the arrow, you could use a ruler to plot the point for the number based on its real and imaginary parts. (The parts specify marching off $100\sqrt{2}$ units along both the x and i axes—that's about 100×1.414, or 141.4 units.) Then you could approximate the real arrow's initial speed by using the ruler to measure the distance between the origin and the point. Alternatively, you could use the venerable Pythagorean theorem to calculate the distance between the origin and the point.

The theorem* states that the sum of the squared lengths of the two shorter sides of any right triangle is equal to the square of the hypotenuse's length. (You might remember this as something like $x^2 + y^2 = z^2$, where x, y, and z stand for the lengths of right triangles' sides.) The right triangle of interest here has two sides of length $100\sqrt{2}$. Its hypotenuse is the arrow, and you can picture one of its sides lying along the x axis. By the Pythagorean theorem, $(100\sqrt{2})^2 + (100\sqrt{2})^2$ equals the square of the hypotenuse. Writing that as an equation and rearranging terms, we have $(100 \times 100 \times \sqrt{2} \times \sqrt{2}) + (100 \times 100 \times \sqrt{2} \times \sqrt{2}) = H^2$, where H is the length of the hypotenuse. That means, using the fact that $\sqrt{2} \times \sqrt{2} = 2$ and further rearranging terms, that H^2 equals $(2 \times 100^2) + (2 \times 100^2) = 2 \times (2 \times 100^2) = (2 \times 100)^2$, which is 200^2. Thus, $H^2 = 200^2$, which implies that $H = 200$.

One conclusion you might draw from all this is that such diagrammed arrows can be thought of as representing the same information as the complex numbers associated with the points at their tips. Thus, it's reasonable to regard the

*This theorem is so famous that it was mentioned in the 1939 movie *The Wizard of Oz*, but not in a way that math teachers would approve of: when awarded a diploma by the wizard, the scarecrow recited an absurdly bungled version of the Pythagorean theorem to demonstrate his newfound braininess.

arrows as visual representations of the associated complex numbers. Such arrow-like complex number representations are called vectors. They're a key component of the geometric interpretation that Wessel pioneered, and, as we'll see, they can be used to extract implications of Euler's formula that even Euler himself never recognized, at least not explicitly.

Notice that when you envision a complex number as a vector, you're translating a two-part number concept into 2-D geometric terms, leading you to picture the abstract number as a visible thing. (Especially if you're an archery fan.) Such visualizations can make it wonderfully easy to carry out basic mathematical operations with complex numbers, effectively translating abstract concepts into concrete images that our brains are naturally equipped to deal with.

In the mid-1800s, Irish mathematician William Rowan Hamilton expanded the concept of numbers even further by introducing four-dimensional numbers called quaternions and working out how to do calculations with them. Such numbers are used today in everything from computer graphics to aircraft navigation systems. Physicists have a thing for many-dimensional numbers too. Einstein pictured the universe as having four dimensions—three spatial ones plus a fourth dimension for time. And infinite-dimensional spaces are used in quantum physics to model properties of elementary particles. (Don't worry if that doesn't mean anything to you—even the White Queen would need a dozen or more pre-breakfast sessions to get her mind around it.)

Thinking of 4-D numbers as vectors would come in handy if you were asked by a Zen master to compose a zero-word autobiography. You could nod sagely and offer a concise depiction of your life to date as a vector in 4-D space. Based on a somewhat arbitrary, earth-centric frame of reference, the vector would extend from a point representing the time and location of your birth (the time might be expressed as estimated number of seconds that had elapsed since Buddha's birth, and the location might be represented by numbers for latitude, longitude, and altitude) to another point similarly representing the current time and your present location. You wouldn't have to figure out how to draw the 4-D vector. Instead, you could effectively delineate it by simply designating its two endpoints in the form (a,b,c,d), where the letters represent the salient numbers for location and time. Of course, this version of your life story would omit all its zigzags and have a very thin narrative element. But the master would like its extreme simplicity. He (or she) might even let you hear the sound of one hand during a standing ovation.

AS SUGGESTED ABOVE, picturing complex numbers as vectors sets the stage for representing addition, multiplication, and other operations with complex numbers in a geometric way. Wessel was the first to work out how to envision such calculations geometrically.

Let's take a look at the geometric interpretation of complex-number addition. Before seeing how such addition is imagined with vectors, however, you should know the rule for doing it arithmetically—that is, how complex numbers are added without reference to vectors. Rafael Bombelli, the Italian mathematician we met in the chapter on imaginary numbers, devised this rule in the 1500s, more than two centuries before the geometric interpretation was introduced. It's easy: you sum two complex numbers by simply adding their real and imaginary parts separately. For example, here's how the complex numbers $3 + 1i$ (which can also be written $3 + i$) and $-1 + 2i$ are added:

$$(3 + 1i) + (-1 + 2i) = (3 + (-1)) + (1i + 2i) = 2 + 3i.$$

The geometric rule for addition of complex numbers entails constructing a parallelogram using the vectors for the numbers as two of its nonparallel sides. A parallelogram is a four-sided figure whose opposite sides are parallel, as shown in Figure 10.3.

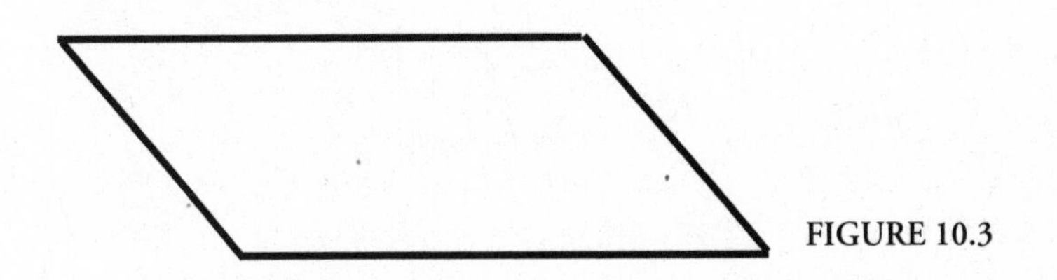

FIGURE 10.3

Drawing a parallelogram to add two vectors effectively defines a third vector that extends diagonally through the parallelogram. This vector represents the sum.

To see what this verbal description means, take a look at Figure 10.4 (next page), which shows the geometric addition of the two complex numbers that were added arithmetically above: $3 + 1i$ and $-1 + 2i$. Note that the summation vector drawn diagonally through the parallelogram—the arrow pointing at $2 + 3i$—represents the complex number that's produced by adding the numbers arithmetically.

FIGURE 10.4

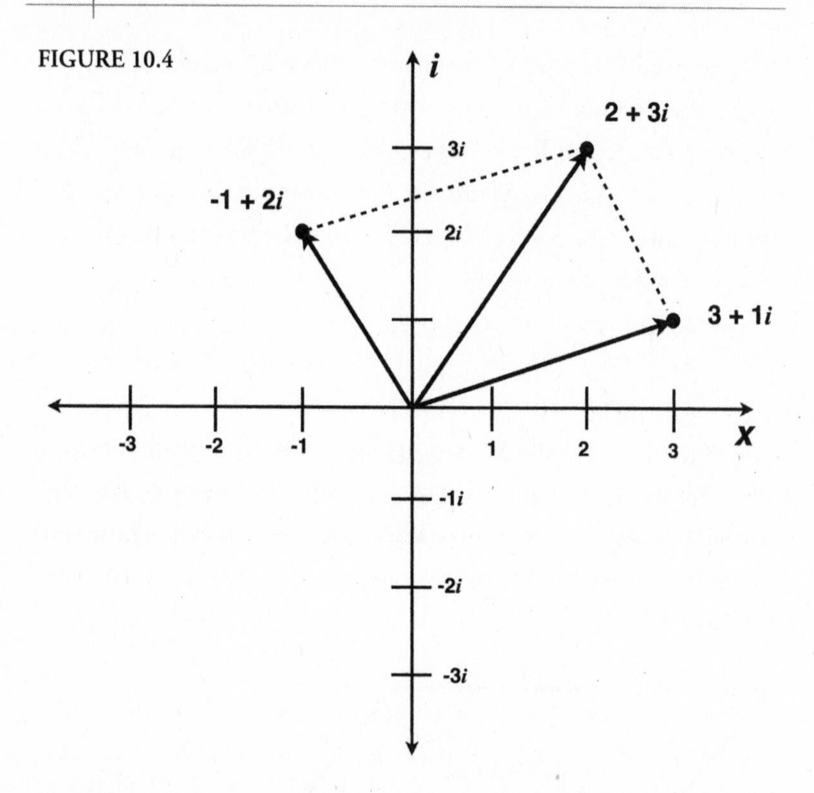

I won't go into all the rules of vector math here; this isn't a textbook. But please take note of an important feature they have in common: they yield results that invariably agree with calculations using complex-number arithmetic, as was just shown with the vector parallelogram for adding $3 + 1i$ and $-1 + 2i$. This consistency between the geometric and non-geometric calculations is critical. If vector-based math yielded something different, the geometric interpretation would be little more than a minor curiosity—like an English translation of *War and Peace* whose sentences were weirdly mangled versions of the original Russian ones. Maintaining strict consistency is crucial in mathematics to prevent its intricate, interconnected towers of logic from collapsing into heaps of contradictions.

Vector addition is intuitively inviting. In particular, it suggests visualizing the addition of complex numbers as the trajectory of an object as it's simultaneously pushed by two forces. (The forces in such visualizations are pictured as the vectors being added together, and the trajectory is the summation vector.)

But the translation of multiplication into vector math is the geometric interpretation's cleverest, most fruitful stroke. In order to keep things simple, I'll focus on a key example of such multiplication: the vector version of multiplying the complex number $0 + i$ times other complex numbers. And I'll lead into that topic with a related example involving only real numbers: the translation of multiplication by -1 into geometric terms.

Since we're accustomed to thinking of real numbers as points on a number line, it's not hard to picture them as vector-like arrows along the number line. Thus, as shown in Figure 10.5, the number 4 can be represented by a four-unit-long arrow extending to the right from 0. (This arrow looks a lot like the vectors defined above, but it's actually a different thing. Vectors in the complex plane are 2-D things, because the plane is a two-dimensional space. The arrow representing 4 is a 1-D thing, because the number line is a one-dimensional space.)

FIGURE 10.5

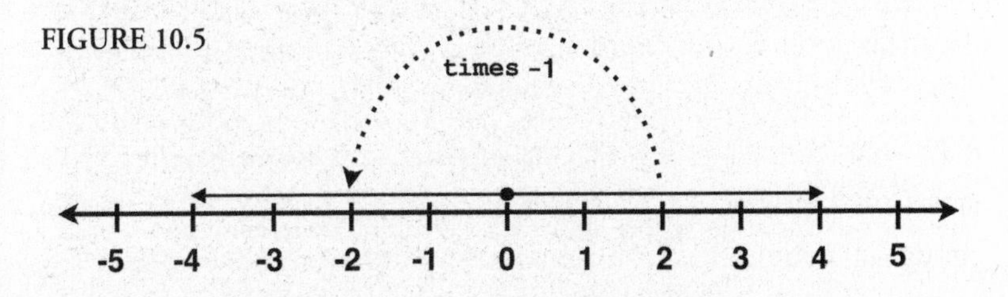

We know from basic arithmetic that $-1 \times 4 = -4$. This idea can be neatly captured in geometric terms by the following operation: when you multiply -1 times 4, you rotate the arrow representing 4 by 180 degrees (with the point for 0 acting as the axis of rotation) to reverse its direction, turning it into the arrow for -4.

Multiplying -1 times the arrow for any real number can be thought of as producing the same kind of 180-degree rotation.* For instance, multiplying -1 times -5 can be pictured as rotating the arrow for -5, which points to the left from 0, by 180 degrees so that it winds up pointing to the right and thus represents 5. (This rotation operation for "times -1," by the way, yields results that always agree with the enemy-of-my-enemy rule for multiplying a negative number times another negative number.)

Now let's extrapolate this rotation idea from the 1-D number line to the 2-D complex plane. To do that, we'll assume that multiplying a complex number times $-1 + 0i$, or -1, is geometrically interpreted as causing the vector for the number to rotate counterclockwise by 180 degrees (with the origin acting as the axis of rotation). If this "times -1" rotation rule works the way it's supposed to, it should yield results that are consistent with those obtained by multiplying complex numbers without the use of vectors.

*In case you hadn't noticed, the idea of rotating a 1-D vector by 180 degrees is actually a bit strange. When you rotate a line segment in your mind as described, it moves off the 1-D number line and pirouettes through 2-D space before returning to the number line. Thus, the concept of rotating a 1-D vector by 180 degrees so that it points backward is something like a 3-D spaceship tunneling through 4-D space to reverse direction. Nonetheless, this sci-fi-like concept works nicely as a geometric representation of multiplying numbers by -1.

Here's an example we can use to check for the desired consistency: multiplying -1 times i. We know from the earlier chapter on imaginary numbers that $-1 \times i$ can be more compactly written as $-i$. Expressed as an equation, that gives us $-1 \times i = -i$, and if we replace the numbers in the equation with the complex-number counterparts, we obtain $(-1 + 0i) \times (0 + i) = 0 - i$.

Now for the vector version, shown in Figure 10.6. As you can see, it involves rotating the vector representing $0 + i$ counterclockwise by 180 degrees in accordance with our rotation rule for the multiplication operation "times -1" (which, recall, is synonymous with the complex-number multiplication operation "times $-1 + 0i$.") This rotation leaves us with the vector representing $0 - i$. Thus, the vector and arithmetical versions of this multiplication are indeed consistent.

FIGURE 10.6

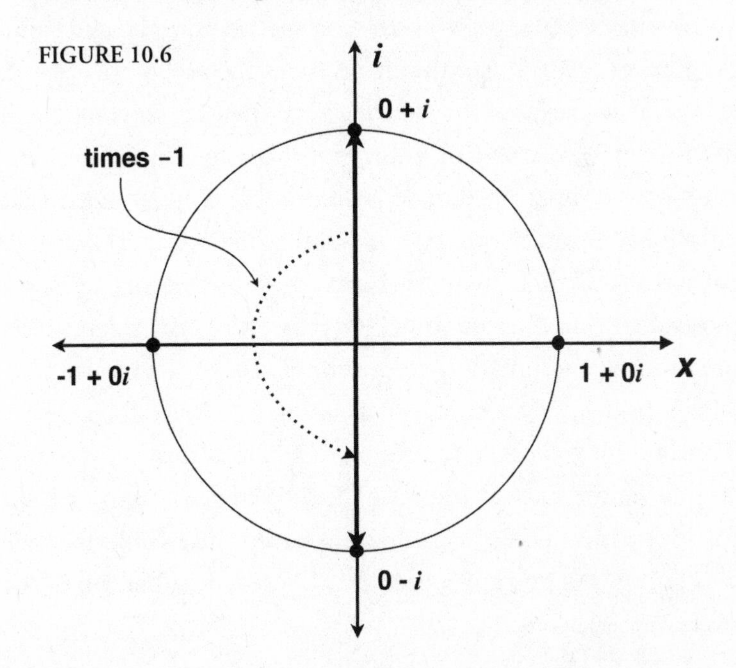

This one example doesn't prove much, but it does suggest that we're on the right track when we interpret complex-number multiplication as vector rotation. And that, in turn, suggests that it may also make sense to geometrically interpret "times i" in terms of vector rotation. Let's consider, for example, how to interpret $i^2 = -1$ along these lines. Expressed as $i \times i = -1$, this true-by-definition equation can be written in complex-number terms as $(0 + i) \times (0 + i) = -1 + 0i$. (Keep in mind that "$(0 + i) \times$" is synonymous with "times i," and that the vector we're multiplying times i in this equation is also $0 + i$, or i.) The equation indicates that when "times i" is applied to the vector for $0 + i$, we should wind up with the vector for $-1 + 0i$. And that implies that "times i" should be geometrically interpreted as inducing a 90-degree counterclockwise rotation. The previous diagram can help you confirm that is true—if you picture rotating the vector for $0 + i$ counterclockwise by 90 degrees, you can see that it becomes the vector for $-1 + 0i$.

But if we assume that "times i" is always associated with a 90-degree vector rotation, will we invariably get vector-based results that are consistent with non-geometric calculations? I don't want to get sidetracked here developing a formal proof that this is the case, or showing lots of examples suggesting that it's true. (Trust me, it is.) But I can't resist mentioning one other case in which the 90-degree rotation rule works like a charm: evaluating i^3, or i cubed. When translated into vector rotations, this calculation represents a simple mathematical model of the old Byrds' song, "Turn! Turn! Turn!" (The song, which is based on a Bible passage, was composed by Pete Seeger. But I best remember the Byrds' version.)

To bring out the turn-turn-turn encapsulated by i^3, it helps to rewrite it as $i \times i \times i \times 1$. This product calls for rotating the vector for 1 (that is, $1 + 0i$) by 90 degrees counterclockwise three times in a row. That leaves the vector pointing in the six o'clock position, which represents the complex number $0 - i$, or $-i$. This checks out with what we know arithmetically, for $i^3 = i \times i \times i = i^2 \times i = -1 \times i = -i$. The triple turn is portrayed in Figure 10.7.

FIGURE 10.7

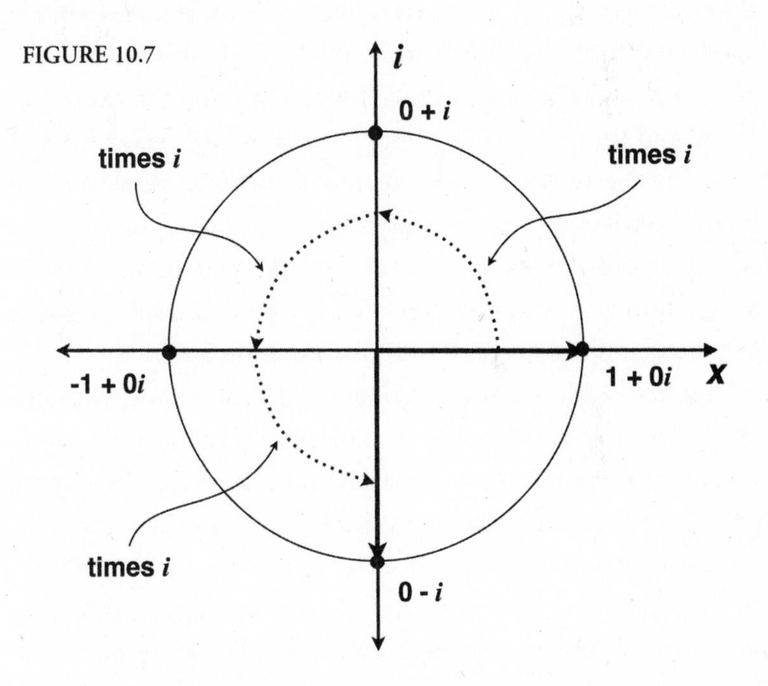

To reiterate, the multiplication operation "times i" is envisioned as a 90-degree counterclockwise vector rotation in the complex plane. Although the line of reasoning I've offered to support this idea differs from Wessel's thinking as he pioneered the geometric interpretation, it leads to the same

90-degree rotation rule for "times i" that follows from his work. The association of multiplication with vector rotation was one of the geometric interpretation's most important elements because it decisively connected the imaginaries with rotary motion. As we'll see, that was a big deal.

AT THIS POINT WE'VE NEARLY covered enough of the geometric interpretation to translate Euler's formula ($e^{i\pi} + 1 = 0$, or $e^{i\pi} = -1$) into vector math and thus look at it in a revealing new way. That is, we know how to represent the constants 1, –1, and 0 as vectors: 1 is represented by the vector for $1 + 0i$, –1 by the vector for $-1 + 0i$, and 0 by an extremely short vector consisting of a single point in the complex plane associated with complex number $0 + 0i$, the origin. We've also covered how to interpret the "+" in Euler's formula in terms of vector addition—recall the parallelogram rule.

But to finish up interpreting the formula in geometric terms, we must confront a more complicated challenge: devising a vector representation of raising a real number to an imaginary-number power. Specifically, we must find a way to represent $e^{i\pi}$ in vector-ese.

Fortunately, we already have a strong clue about what raising e to an imaginary-number power should do in vector math: It should yield results that are consistent with those obtained by raising e to imaginary-number powers in the equation $e^{i\theta} = \cos\theta + i\sin\theta$. (We know this equation is true based on Euler's three non-geometric derivations.) In other words, the vector for $e^{i\theta}$ should be interpreted in such a way that it always turns out to be the vector for $\cos\theta + i\sin\theta$ when an angle—any angle—expressed in radians is plugged in for θ in these expressions. If that weren't the case, the consistency between geometric and non-geometric calculations that has worked so nicely for us as a guiding principle would be fatally compromised.

It follows that if we can figure out how to translate $\cos\theta + i\sin\theta$ into the visual language of vectors, we'll also know how $e^{i\theta}$ should be translated. Fortunately, we've already gone over most of the concepts needed to solve this problem.

In Figure 10.8 (next page), I've reformulated an illustration from the trigonometry chapter to help you see how to apply what you already know about trig to the problem at hand. It depicts adding the vector representing $\cos\theta + 0i$, which is a pure real complex number (since $\cos\theta$ is a real number), to the vector for i times $\sin\theta$, which is a pure imaginary complex number,* using the parallelogram rule for vector addition. (In this case, the parallelogram's sides meet at right angles, making it a rectangle.)

The summation vector represents the complex number $\cos\theta + i\sin\theta$. This vector sum, by the way, is consistent with the sum obtained by adding the complex numbers arithmetically:

$$(\cos\theta + 0i) + (0 + i\sin\theta) = (\cos\theta + 0) + (0i + i\sin\theta)$$
$$= \cos\theta + i\sin\theta.$$

---

*The sine of θ, or $\sin\theta$, is a real number, just as is $\cos\theta$. Thus, a vector representing $\sin\theta$ by itself would lie along the x axis in the complex plane, as do the vectors for all pure real complex numbers. But as we saw earlier in this chapter, multiplying a complex number times i effectively makes the vector representing that number rotate by 90 degrees. Therefore, multiplying i times $\sin\theta$ (written $i\sin\theta$), when interpreted geometrically, rotates the x-axis-aligned vector for $\sin\theta$ by 90 degrees so that it winds up aligned with the i axis. In fact, $i\sin\theta$ can be pictured as a vector lying right on top of the i axis.

FIGURE 10.8

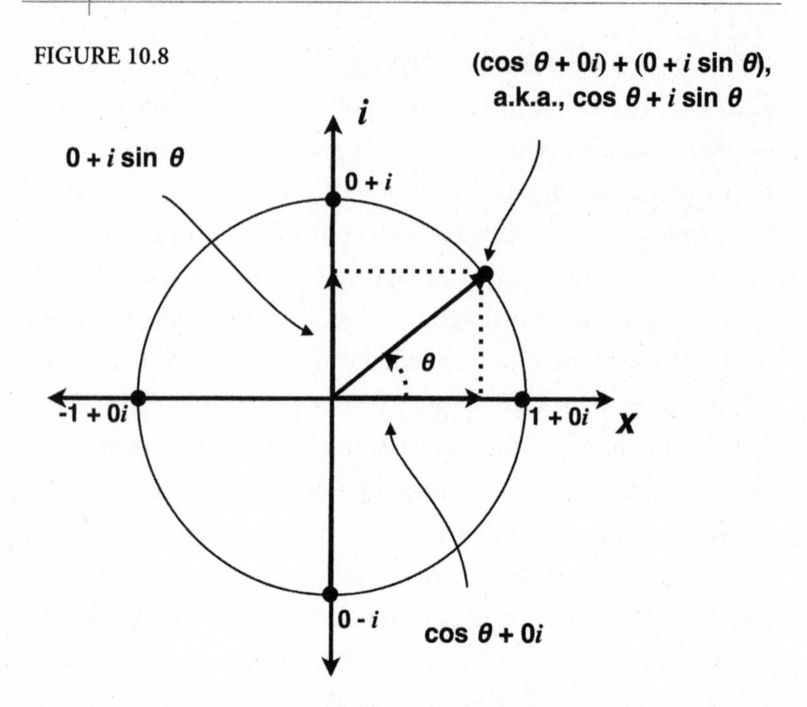

As you can see in the diagram, the vector for cos θ + i sin θ closely resembles the angle sweeper envisioned in the trig chapter. Remember that the coordinate pair of the point at the sweeper's tip is (cos θ, sin θ). As you may have already realized, cos θ + i sin θ is the complex-plane version of this coordinate pair. The definitions of the unit-circle-based trig functions guarantee that the point associated with the complex number cos θ + i sin θ will lie on the unit circle in the complex plane, just as (cos θ, sin θ) always does on the unit circle in the xy plane. Further, it will be located at a horizontal distance of cos θ from the i axis, and at a vertical distance of sin θ from the x axis.

Borrowing another idea from the trig chapter, we can think of cos θ + i sin θ as a sort of device for controlling a radius-like angle-sweeping vector, as suggested in the previous diagram. And in keeping with the angle-sweeper action pictured in the trig chapter, when a number (expressed in radians) is plugged in for θ in $\cos \theta + i \sin \theta$, the angle-sweeping vector can be envisioned as rotating counterclockwise from the three o'clock position by that number of radians.

Don't look now (actually, do look—at Figure 10.9), but we've just arrived at a plausible geometric interpretation of $e^{i\theta}$. To wit: *$e^{i\theta}$ can be geometrically interpreted as an angle-sweeping vector that rotates within the unit circle in the complex plane to sweep out the angle θ.*

The geometric interpretation of $e^{i\theta}$ represents a wonderfully concise, convenient way to represent the sweeping out of angles. Notice how the trig-function clutter has been dis-

FIGURE 10.9

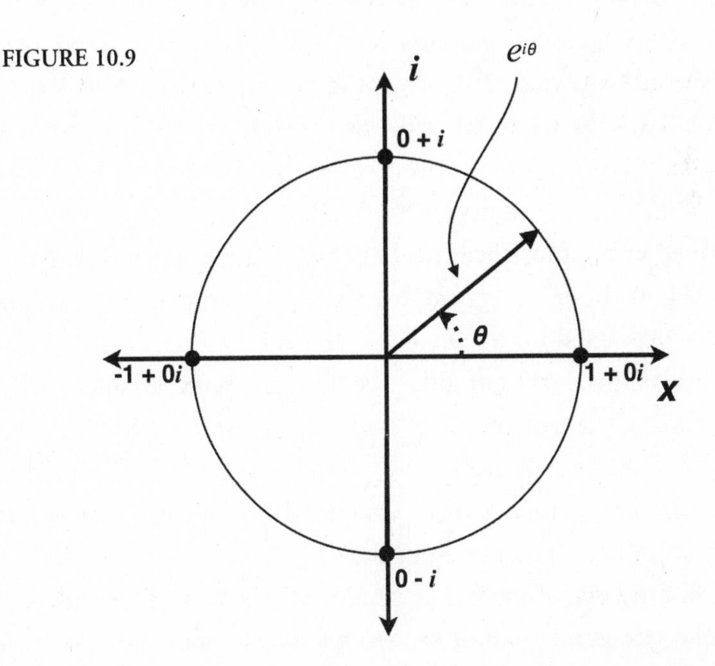

pensed with in the diagram because it's no longer needed, conceptually speaking.

Now let's imagine what happens when we plug in $\pi/2$ for the θ in $e^{i\theta}$. Since $\pi/2$ radians is the same as 90 degrees, the vector version of $e^{i\pi/2}$ can be interpreted as the angle-sweeping vector after it has carried out a 90-degree counterclockwise rotation, from the three o'clock to the high-noon positions. That means it winds up representing the complex number $0 + i$. To check that this vector operation is correct, we can plug $\pi/2$ into $\cos\theta + i\sin\theta$ (which, remember, is equivalent to $e^{i\theta}$) and treat it as a complex number. Since $\cos \pi/2 = 0$ and $\sin \pi/2 = 1$, we have $\cos \pi/2 + i \sin \pi/2 = 0 + (i \times 1)$, or $0 + i$. That, of course, is what we wanted to see.

Similarly, when π radians is plugged into $e^{i\theta}$, the angle-sweeping vector representing it rotates counterclockwise by 180 degrees (π radians) to wind up in the nine o'clock position. Note that after sweeping out π radians, the $e^{i\theta}$-based angle sweeper winds up representing $-1 + 0i$. If we translate this statement into one about complex numbers, we have $e^{i\pi} = -1 + 0i$, or, more simply, $e^{i\pi} = -1$. In other words, we've just arrived at the most beautiful equation in a geometric way.

What about the geometric interpretation of Euler's formula expressed in the usual way, $e^{i\pi} + 1 = 0$?

This equation's left side can be interpreted geometrically as adding the vector for $e^{i\pi}$, which we established above is identical to the vector for $-1 + 0i$, to the vector for $1 + 0i$. This vector sum, in turn, can be imagined as the trajectory of an object, initially positioned at the origin, as it's simultaneously pushed by two forces. (As noted earlier, this force-pushing analogy for vector addition is suggested by the parallelogram rule for summing vectors.) Since the vectors being summed in this case resemble equal-length arrows pointing in opposite directions, the hypothetical dual forces pushing on the object exactly cancel each other out, resulting in a trajectory consisting of a non-moving point, the origin. And the origin, of course, is $0 + 0i$—the complex-number version of the pure real number 0, which is the equation's right side.

So here's the main message to take home from this chapter: *Raising* e *to an imaginary-number power can be pictured as a rotation operation in the complex plane. Applying this interpretation to* e *raised to the* "i *times* π" *power means that Euler's formula can be pictured in geometric terms as modeling a half-circle rotation.*

BEFORE MOVING ON to the implications of the take-home message, which is covered in the next and last chapter, let's briefly consider how the geometric interpretation enables mental shortcuts.

As we just saw, plugging in $\pi/2$ radians for θ in $e^{i\theta}$ can be pictured as making a three-o'clock-pointing vector within the unit circle rotate counterclockwise by 90 degrees, and thus $e^{i\pi/2} = 0 + i$, or, more concisely, $e^{i\pi/2} = i$. Notice that the geometric interpretation immediately handed us this equation without our having to fiddle with trig functions or carry out other math machinations. This conceptual efficiency is due to the fact that we're now thinking about 2-D things (complex numbers) that have been reformulated as vectors that we can simply rotate in our minds to perform calculations. This mental efficiency has helped make $e^{i\theta}$ very useful in engineering and science.

Now for a last variation on the theme. Let's plug in 2π radians for θ in $e^{i\theta}$, causing the sweeper to rotate by 360 degrees and wind up back where it started, in the three o'clock position pointing at $1 + 0i$. That gives us

$$e^{i2\pi} = 1 + 0i$$

which, by subtracting 1 from both sides and using 0 for $0 + 0i$, becomes

$$e^{i2\pi} - 1 = 0.$$

This equation has been known as long as Euler's formula, and although it doesn't have the cachet of $e^{i\pi} + 1 = 0$, I regard it as suitable for framing. It features all five of the very important numbers seen in the famed formula as well as another VIN: the number 2, "a couple," the fundamental number for romance. I like to call it Alicia's formula, after my wife, a former mathphobe now in recovery. (She kindly served as a test reader of this book as it was written.)

THE GEOMETRIC INTERPRETATION of $e^{i\pi}$ is rich with emblematic potential. You could see its suggestion of a 180-degree spin as standing for a soldier's about-face, a ballet dancer's half pirouette, a turnaround jump shot, the movement of someone setting out on a long journey who looks back to wave farewell, the motion of the sun from dawn to dusk, the changing of the seasons from winter to summer, the turning of the tide. You could also associate it with turning the tables on someone, a reversal of fortune, turning one's life around, the transformation of Dr. Jekyll into Mr. Hyde (and vice versa), the pivoting away from loss or regret to face the future, the ugly duckling becoming a beauty, drought giving way to rain. You might even interpret its highlighting of opposites as an allusion to elemental dualities—shadow and light, birth and death, yin and yang. Math historians Edward Kasner and James R. Newman once observed that Euler's formula "appeals equally to the mystic, the scientist, the philosopher, the mathematician." It seemingly might also appeal to those with a poetic turn of mind, for it suggests that when three of the most elemental numbers are combined, they somehow spring to life and speak of ducklings and dancers, transformations and farewells.

CHAPTER 11

The Meaning of It All

By the early 1800s several mathematicians, including Gauss, had independently developed the idea of geometrically representing complex numbers. Still, it was a major leap, and such leaps are generally far from obvious before they're made. Indeed, it had eluded even Euler's grasp. Although he was familiar with the idea of vectors, there's no evidence that he visualized complex numbers as vectors that could be manipulated on a 2-D plane to represent calculations.

One indicator of the advance's importance is that it blew away the air of impossibility that had long surrounded imaginary numbers. In effect, Wessel and his fellow explorers had discovered the natural habitat of Leibniz's ghostly amphibians: the complex plane. Once the imaginaries were pictured there, it became clear that their meaning could be anchored to a familiar thing—sideways or rotary motion—giving them an ontological heft they'd never had before. Their association with rotation also meant that they could be conceptually tied to another familiar idea: oscillation. Eventually the formerly confusing will-o'-the-wisps came to be seen as solid players in physics and engineering for, among other things, representing phenomena that involve regular, back-and-forth patterns. (One such phenomenon is especially close to home: our very bodies oscillate daily via circadian rhythms.)

Pioneering electrical engineer Charles Proteus Steinmetz spearheaded the use of the imaginaries in calculations related to alternating current.* A diminutive hunchback, he'd fled his native Prussia as a young man and emigrated to America after being threatened with arrest for supporting socialist causes. Just months after arriving in the United States he began making fundamental advances that revolutionized the use of electricity. In 1892 he joined the newly formed General Electric and, soon after, published a landmark paper showing how to use imaginary numbers to greatly simplify analysis of AC circuits.

Delighted by animals that were typically shunned, Steinmetz kept a menagerie of peculiar pets in his Schenectady, New York, mansion, including alligators, rattlesnakes, and black widow spiders. School improvement was another of his avocations, and he pushed for the introduction of special classes for immigrants' children. Once when Thomas Edison visited him, he delighted the aged, nearly deaf inventor by tapping out messages on

---

*Alternating current is electricity that reverses direction many times a second, giving it the character of a pendulum that rapidly swings back and forth—it oscillates. AC's voltage can be readily ramped up and down by transformers, like the ones cased in metal boxes on telephone poles. That enables its efficient long-distance transmission from generators at very high voltage; the voltage is then reduced for relatively safe home use by local transformers. AC's ability to be efficiently transmitted all over the place is why it is used in the electric grid instead of direct current, the unidirectional electricity that flows from batteries.

Edison's knee in Morse code. In later life he was dubbed "The Wizard of Schenectady" in the media, and at some point a more elaborate epithet was invented: "The Wizard Who Generated Electricity from the Square Root of Minus One."

Euler's general formula, $e^{i\theta} = \cos \theta + i \sin \theta$, also played a role in bringing about the happy ending of the imaginaries' ugly duckling story. Even before the geometric interpretation shed new light on imaginary numbers, Euler worked out some remarkable things about them based on the formula. An example is the evaluation of i^i, shown in Appendix 2.

The interpretation of $e^{i\theta}$ as a rotating vector paved the way for constructing particularly elegant mathematical models of rotation and oscillation using this compact function. Such "exponential models" make it surprisingly easy to carry out calculations that, in many cases, would be considerably more difficult if trigonometric functions, the main alternative, were used instead. Recall from the last chapter how simple it was to derive $e^{i\pi/2} = i$ by thinking of $e^{i\theta}$ as an angle sweeper and mentally rotating it. The use of exponential models also brings into play the ease of applying calculus to e^x-based functions, mentioned in Chapter 2—$e^{i\theta}$ offers a very similar user-friendliness.

Today, Euler's formula is a tool as basic to electrical engineers and physicists as the spatula is to short-order cooks. It's arguable that the formula's ability to simplify the design and analysis of circuits contributed to the accelerating pace of electrical innovation during the twentieth century.

EULER'S GENERAL EQUATION stands out because it forged a fundamental link between different areas of math, and because of its versatility in applied mathematics. After Euler's time it came to be regarded as a cornerstone in "complex analysis," a fertile branch of mathematics concerned with functions whose variables stand for complex numbers.

But the special case, $e^{i\pi} + 1 = 0$, is mainly treasured because it's beautiful. What makes it as exquisite as a great poem or painting?

I doubt that there's a simple answer to this question that most of those who find the formula beautiful would agree with. In fact, some math lovers don't regard it as particularly beautiful, nor do they even find it very interesting—more on that momentarily—which goes to show that the eye-of-the-beholder issue will always arise regardless of whether we're contemplating art or mathematics. (This is why aesthetics strikes me as both endlessly provocative and fundamentally absurd.) But despite the risk of getting mired in a morass of conflicting opinions, I feel obliged to address the beauty question—after all, the word beauty is in the title.

First, let me frame what I'm calling beautiful. It's not simply the equation's neat little string of symbols. Rather, it's the entire nimbus of meaning surrounding the formula, including its funneling of many concepts into a statement of stunning brevity, its arresting combination of apparent simplicity and hidden complexity, the way its derivation bridges disparate topics in mathematics, and the fact that it's rich with implications, some of which weren't apparent until many years after it was proved to be true. I think most mathematicians would agree that the equation's beauty concerns something like this nimbus.

But what makes the nimbus beautiful?

Dictionaries define beauty as qualities that give pleasure or deep satisfaction to the senses or the mind. That's nice, but it only leads to another question: Why does Euler's formula induce pleasure?

A good starting point on that is a much-cited observation by the great British philosopher and mathematician Bertrand Russell:

> Mathematics, rightly viewed, possesses a beauty cold and austere, like that of sculpture, without appeal to any part of our weaker nature, without the gorgeous trappings of painting or music, yet sublimely pure, and capable of a stern perfection such as only the greatest art can show. The true spirit of delight, the exaltation, the sense of being more than Man, which is the touchstone of the highest excellence, is to be found in mathematics as surely as poetry.

There's ample food for thought in this eloquent declaration, but the first sentence bothers me. Russell's reference to a cold, austere, sternly perfect quality that doesn't appeal to our weaker nature strikes me as reinforcing sadly common negative stereotypes about math: that it's dry, forbidding, and rock-hard. Further, he seems to be denigrating what might be called the soft power of music, painting, and other arts to elicit feelings—the gorgeous trappings by which they presumably enchant us. That is, he's implying that the Dionysian realm is a lesser thing than the stricter, purer Apollonian realm, and that only the latter is associated with the beauty of mathematics and the greatest art.

It's true that feelings are gorgeously messy and inextricably linked with subjectivity, which is seemingly the opposite of math's stern objectivity. But it is also true that the limbic system (the brain system most closely associated with emotions) is crucial to the way our minds work. I suspect that if its activity were somehow greatly dialed down so that a person could experience genuinely cold, austere pleasure when contemplating, say, Euler's formula, the result would be a mental state resembling that of a weirdly robotic Spock.*

Thus, while feelings may be the essence of subjectivity, they are by no means part of our weaker nature—the valences they automatically generate are integral to our thought processes and without them we'd simply be lost. In particular, we'd have no sense of beauty at all, much less be able to feel (there's that word again) that we're in the presence of beauty when contemplating a work such as Euler's formula. After all, $e^{i\pi} + 1 = 0$ can give people limbic-triggered goosebumps when they first peer with understanding into its depths. (I'm one.)

---

*Actually, I think that Spock, for all his stern, austere logic, had a perfectly functional, human-like limbic system. (After all, he was half human—his father was Vulcan.) Note that he had little trouble interacting with the earthlings around him. Thus, despite his ostensible lack of emotions, he was not at all like a severely autistic person who was essentially flying blind when it came to others', as well as his own, feelings, and who therefore could barely operate in society. Indeed, my sense is that Spock had a secretly high emotional intelligence, and his fastidious pretense of being an affective doofus was exactly what made him so charismatic. Whenever he was in a scene, I, for one, found him upstaging the other Star Trek characters, and it wasn't simply because of his pointy ears, amused ironic wit, or precise manner of speaking—it was largely because, as a supposedly very-low-emotional-intelligence being who was paradoxically competent in social interactions, he enigmatically defied my instinctive sense of what is psychologically possible.

But what accounts for this limbic-mediated thrill? I think it springs from a combination of things, including the equation's seriousness, generality, depth, unexpectedness, inevitability, and economy—qualities that prominent twentieth-century English mathematician G. H. Hardy singled out as key ingredients of mathematical beauty. (Hardy was much concerned with the nature of mathematical beauty because he held that "there is no permanent place in the world for ugly mathematics." That isn't quite as strong as the celebrated Keats formula, beauty = truth, but it's pretty close to it.)

There's also the formula's elegance, a word that mathematicians use to designate a clever mix of concision and sophistication. And then there's the cool way that disparate ideas cunningly fit together in its derivation—we get essentially the same kick from this aspect of mathematical beauty that people get when they puzzle out how to snap together elaborate Lego windmills or spaceships at a certain age (like 46, when the kid gets a big Lego set for Christmas). *Homo mechanicus* are us.

But I think Euler's formula most importantly invokes a rarer, deeper thrill: the feeling of exaltation that we get from an encounter with an example of our species outdoing itself. I get pretty much the same thrill from the creations of Michelangelo, Beethoven, Jane Austen, George Eliot, and W. H. Auden; from the discoveries of Charles Darwin, Marie Curie, and Albert Einstein; from Lincoln's managing to hold the union together while ending slavery in the United States; from Helen Keller's incredible achievements; from the monumental persistence of Elizabeth Cady Stanton, Susan B. Anthony, and the other marathoners of the spirit who secured basic rights for women; and from Nelson Mandela's world-changing magnanimity. This is what Russell's second sentence is about, and he put it beautifully despite his outdated use of "Man" to refer to all of us—his phrase "the sense of being more" really nails it.

What I'm getting at here is closer to what's called the sublime in aesthetic theory than to beauty. Ideas about the sublime go back to Longinus, a first-century Greek thinker who described it in terms of grand, awe-inspiring thoughts or words. Later thinkers posited that the sublime and the beautiful are different; the former is supposedly associated with feelings of horror-tinged awe, such as those felt by an astronaut looking at the receding earth while hurtling toward the moon, while beauty is mainly about pleasure. While this distinction is interesting, what I find most salient is the idea of exaltation from the sense of being more.

Chinese philosopher Tsang Lap-Chuen is a leading modern exponent of the idea that the sublime involves this kind of experience. In *The Sublime: Groundwork towards a Theory*, published in 1998, he wrote that the sublime "evokes our awareness of our being on the threshold from the human to that which transcends the human; which borders on the possible and the impossible; the knowable and the unknowable; the meaningful and the fortuitous; the finite and the infinite." In his view, there is no single essential common property possessed by sublime works or sublime natural objects, nor is there a single emotional state evoked by all of them. But he argues that there's a common thread in experiences of the sublime, which is that they take us "to the limit of some human possibility."

Euler's formula may seem elementary to modern mathematicians, but many of them still feel that it's extraordinarily beautiful. I think this may largely be because they've retained a lively sense of it as emblematic of the sense of being more—it represents the true story of how an all-but-supernatural genius reached beyond what once seemed possible to come up with a deep, almost miraculously concise truth. Thus, their familiarity with it breeds no contempt. To them, as to me, Euler's formula is a joy forever.

This exalting, sublime-related kind of beauty is relatively rare, and of course the word beautiful applies to other sorts of things. But as Russell noted, great math and great art both possess it. And that points to a longstanding conundrum in aesthetics: How can it be that certain works are prized as beautiful/sublime across many generations, somehow defying the incessant zigzags of fashion and the continual rethinking of what's deeply satisfying to the senses or mind? A number of Paleolithic cave drawings, for example, are widely regarded as sublimely beautiful some 30,000 years after their creation—I've seen some of them up close in France and can testify that they are hair-raisingly, impossibly superb. Euler's formula has similarly retained its charismatic appeal to mathematicians across generations. The brain-scan study that was mentioned in the first chapter, involving mathematicians' neural responses to certain equations, suggests why: the durability of beauty in mathematics and other realms is based, at least in some cases, on ubiquitous aspects of the human mind. Beauty may be in the brain of the beholder, but beholders' brains (including their limbic systems) seem to react similarly when confronted with truly sublime rarities, such as Euler's formula.

AND YET SOME PEOPLE assert that Euler's formula is much overrated. It's obvious, they say—there's no mystery, no paradox, no reaching into the depths of existence.

I've seen variations on this theme in a number of online commentaries. One blogger even claimed that Euler's formula is so simple that "small children" understand its meaning. French chemical engineer, writer, and amateur mathematician François Le Lionnais struck a similarly jaded stance: the formula "seems, if not insipid, at least entirely natural," he wrote. One skeptical respondent in the 1988 survey that ranked Euler's formula as mathematics' most beautiful result commented that it's "too simple" to classify as supremely beautiful, and another apparently gave it a low rating because it's simply "true by virtue of the definition" of its terms.

My take on these opinions probably won't surprise you—they strike me as abusing hindsight.

True, Euler's equation lacks the charm of novelty or the charisma of a major unanswered question in mathematics. And its derivation is straightforward compared to, say, the celebrated proof of Fermat's Last Theorem by mathematician Andrew Wiles—published in 1995, it required over 150 pages of complex mathematics.* But I think those who assert that Euler's formula is obvious, or that the connections it reveals are less than amazing, are displaying a lack of historical perspective as well as a sadly blunted sense of wonder. They seem

*In 1637 French mathematician Pierre de Fermat conjectured that no positive integers a, b, and c satisfy the following equation (make it true) if each number's exponent, n, is a whole number greater than 2: $a^n + b^n = c^n$. This became known as Fermat's Last Theorem, and mathematicians tried to prove it for 357 years before Wiles finally did so. For $n=1$ or $n=2$, however, it's easy to find values of a, b, and c that work. For instance, $3^2 + 4^2 = 5^2$.

to believe that understanding and wonder are fundamentally incompatible, and that people who marvel at Euler's equation are either pretentious or mathematically naïve. The same logic would imply that marveling at Michelangelo's *David* is incompatible with understanding exactly how it was carved out of marble.

Perhaps those who assert that the formula is borderline boring are implicitly claiming that after you've mastered the main math concepts that are combined in the equation, and know how to derive it, you should find it no more interesting or profound than $2 + 2 = 4$. To me, however, this would be akin to a champion hurdler declaring, "Hey people, leaping over a bunch of three-and-a-half-foot hurdles while sprinting as fast as you can is really no big deal. I can show you how ridiculously simple it is in nothing flat. Even small children can do it."

It took mathematicians centuries to make the conceptual advances that led to the equation and its geometric interpretation. These ideas were unknown to many brilliant mathematicians before the nineteenth century. The reason is clear: they aren't obvious, period. And understanding how to derive the formula only highlights further loci of wonder. Consider again $e^{\theta} = 1 + \theta + \theta^2/2! + \theta^3/3! + \cdots$. When 1 is plugged in for θ, the equation becomes $e = 1 + 1 + 1/2! + 1/3! + \cdots$, which can be rewritten as $e = 1/0! + 1/1! + 1/2! + 1/3! + \cdots$ (since by definition the factorials 0! and 1! both equal 1). Despite knowing how to derive this last equation from basic principles, I still find it beautiful and surprising. It shows that e, which at first glance seems to be a messy, endlessly chaotic irrational number (based on its definition as the limit of $(1 + 1/n)^n$ as n goes to infinity), is a perfect paradigm of order and simplicity—in fact, just a slightly dressed-up version of 0, 1, 2, 3,...—when viewed from another perspective.

Keith Devlin, the mathematician who compared Euler's formula to a Shakespearean sonnet, has summed up what such surprises seem to be telling us. "Surely," he wrote, such a formula "cannot be a mere accident; rather, we must be catching a glimpse of a rich, complicated, and highly abstract mathematical pattern that for the most part lies hidden from our view." Those who say that Euler's formula is ho-hum imply that they can perceive the whole of this pattern and all of its implications. Perhaps they really are incredible geniuses and have managed to do that. I've not seen convincing evidence that they have.

But my main objection to the killjoy view of Euler's formula has to do with a practical matter. The most inspiring teachers I've known possessed the gift of infectious enthusiasm—they communicated intellectual excitement about their subjects by seeming to regard them with the fresh eyes of impassioned novices. While explaining things, they often appeared to be re-experiencing, or at least to be re-enacting, what it was like when they first fell in love with their specialties. I've tried to do the same thing in this book. The killjoys dismiss such approaches as simple-minded. While there's no disputing taste, I must confess that I wouldn't want them as my kids' math teachers. I'm convinced that mathematics seems boring to many people largely because they've never caught on to the fact that it's full of beauty and surprise. The killjoys seemingly want to keep that under wraps, as if to prove that rigor = *rigor mortis*.

I SUSPECT THAT INTELLIGENT BEINGS on a distant planet would discover many of the number- and logic-based relationships contained in our math textbooks, perhaps including Euler's formula. This belief reflects my view that mathematics is grounded in patterns that exist independently of our minds and thus is concerned with objective truths. This intuition has helped shape my thinking about Euler's formula. In a way it has been the elephant in the room throughout this book.

The idea that statements like $1 + 1 = 2$ and $e^{i\pi} + 1 = 0$ express truths that exist independently of human thought is called mathematical Platonism. G. H. Hardy was one of the most prominent modern math Platonists. In an essay on his life's work, he wrote, "I believe that mathematical reality lies outside us, that our function is to discover or observe it, and that the theorems which we prove, and which we describe grandiloquently as our 'creations,' are simply our notes of our observations. This view has been held, in one form or another, by many philosophers of high reputation from Plato onward...."

While I lean toward a form of Platonism, Hardy's purist version doesn't appeal to me. Harvard mathematician Barry Mazur nicely described the kind of ambivalence I experience on this topic in his book *Imagining Numbers (Particularly the Square Root of Minus Fifteen)*: "On the days when the world of mathematics seems unpermissive, with its gem-hard exigencies, we all become fervid Platonists (mathematical objects are 'out there,' waiting to be discovered—or not) and mathematics is all discovery. And on other days, when we see someone who...seemingly by willpower alone, extends the range of our mathematical intuition, the freeness and open permissiveness of mathematical invention dazzle us, and mathematics is all invention."

Such ambivalence is seldom voiced, however, and debates about math's basic nature have long been a philosophical version of the sweet science, replete with lots of complicated bobbing, weaving, and punching. Sometimes the fights have even spilled over into the pages of nontechnical publications. For instance, two celebrated heavyweights—math popularizer Martin Gardner and mathematician Reuben Hersh—clashed about it in such venues as *The New York Review of Books* in the 1980s and 90s. Gardner, who died in 2010, was a mathematical realist (math realism is basically the same as Platonism). He argued that "if all intelligent minds in the universe disappeared, the universe would still have a mathematical structure, and that in some sense even the theorems of pure mathematics would continue to be 'true.'"

Hersh, also a very distinguished writer (his 1981 book, *The Mathematical Experience*, co-authored with mathematician Philip J. Davis, won a National Book Award), countered that mathematics is a human cultural construct that has no reality independent of people's minds. Its statements are invented "social objects" like institutions and laws, in his view. Thus, it's wrong to speak of math as true in any timeless sense—even statements such as 2 + 2 = 4 lack infallibility. And although he says that mathematics is objective, he interprets the word objective to mean "agreed upon by all qualified people who check it out"—not "out there" in some sense. "Saying [mathematics] is really 'out there,'" he adds, "is a reach for a superhuman certainty that is not attained by any human activity."

There is much to commend in Hersh's writing. His 1997 book, for example, includes a highly readable account of the history of thinking about the basic nature of mathematics. But I doubt that he and other advocates of "social constructivist" theories about mathematics have won over many Platonists. In a 1997 review of such theories, for instance, Gardner ardently reiterated his opinion that "the world out there, the world not made by us, is not an undifferentiated fog. It contains supremely intricate and beautiful mathematical patterns.... It takes enormous hubris to insist that these patterns have no mathematical properties until humans invent mathematics and apply it to the world."

In 2000, cognitive scientists George Lakoff and Rafael E. Núñez launched another provocative attack on mathematical Platonism in their book *Where Mathematics Comes From: How the Embodied Mind Brings Mathematics into Being*. They acknowledge that Hersh "has long been one of [their] heroes." But they disagree with the most radical implications of social constructivist theories, emphatically stating that they themselves "are *not* adopting a postmodernist philosophy that says that mathematics is *merely a cultural artifact*." (Emphasis in the original.)

Lakoff and Núñez posit that mathematics—or, at least, basic arithmetic—is grounded in our sensory and motor experiences. For instance, as children we map the idea of a measuring stick's length to the size of a number. Such "grounding metaphors" anchor basic math in the objectively knowable, real world, which includes our neural and other bodily apparatus—thus, their term "embodied" mathematics. In their view, "Two plus two is always four, regardless of culture," because such statements are based on real-world objects, which possess culture-independent qualities such as stability, consistency, generalizability, and discoverability.

More complex mathematical ideas, such as Euler's formula, are based on "conceptual metaphors" that link and blend math concepts, according to their theory. Conceptualizing the sine and cosine functions as infinite sums based on terms like $\theta^n/n!$, for instance, entails blending conceptual metaphors about functions as numbers, numbers as wholes that are the sum of their parts, and limits, they explain.

They argue that math Platonists are wrong because mathematics mainly consists of layers upon layers of conceptual metaphors that exist only in human minds. In their view, the belief that math possesses some sort of "transcendent" reality is a mystical doctrine with no empirical support. They further assert that mathematical Platonism is central to an elitist culture in math that "rewards incomprehensibility" and "has contributed to the lack of adequate mathematics training in the populace in general." (See what I mean about the throwing of punches?)

THE SURPRISING CONNECTIONS that frequently crop up in mathematics bear on this debate. Do they show that mathematical patterns exist independently of our minds, and that we sometimes only gradually perceive and piece them together into a gestalt—like flashes of reflected light seen through clouds from an airplane that turn out to be Lake

Erie? Or do they spring from the fact that mathematicians sometimes invent conceptual metaphors that are so intricately interlinked that it often takes a long time for people to see all the implied connections?

And what about the many math advances that have seemed at first to be no more than the products of an abstract, rule-based game with no bearing on the real world, and then later have clicked into place as amazingly well-suited for representing physical phenomena? Euler's general formula is an example—long after Euler derived it, engineers found it to be just what they needed to help model AC circuits. Such uncanny coincidences often resemble fantasy-novel plot twists: "Suddenly Sam realized that the mysterious ornament he'd taken from the ancient Egyptian mummy was actually a key for unlocking the door leading to the hyperconium reactor's control room."

Physicist Eugene Wigner famously dubbed this phenomenon "the unreasonable effectiveness of mathematics in the natural sciences." On my Platonist-leaning days, I see this effectiveness as suggesting that mathematics is out there, and true independently of us. Not surprisingly, Lakoff and Núñez disagree: "Whatever 'fit' there is between mathematics and the world," they aver in *Where Mathematics Comes From*, "occurs in the minds of scientists who have observed the world closely, learned the appropriate mathematics well (or invented it), and fit them together (often effectively) using their all-too-human minds and brains."

In a review of Lakoff and Núñez's book, mathematician and author John Allen Paulos, who wrote the bestseller *Innumeracy: Mathematical Illiteracy and Its Consequences* (1988) and other entertaining books, offered a "quasi-Platonist" compromise that I find appealing. "Arithmetic may...be transcendent in the sense that any sentient being would eventually develop the metaphors that ground it and be led to its truths, which can thus be said to inhere in the universe," he wrote. (Such hypothetical beings have come up frequently in the debate—Hersh, for example, has allowed that "little green critters from Quasar X9" may do mathematics, but has argued that their math could well be totally different from ours.)

I'm drawn to Paulos's quasi-Platonism largely because it fits with my hunch that the brains of far-away sentient beings, if they exist, would probably have been shaped by an evolutionary process that works like the one that gave rise to human intelligence. Basic quantitative and abstracting abilities could confer a Darwinian edge in many situations. For instance, such faculties would be invaluable for beings that band together in environments with limited resources, and that exchange goods and services. Based on this logic, evolutionary thinker Haim Ofek has theorized that resource exchanges helped drive the explosive growth in brain size and cognitive abilities that led to modern humans. As he has observed, "Exchange requires certain levels of dexterity in communication, quantification, abstraction, and orientation in time and space—all of which depend (i.e. put selective pressure) on the lingual, mathematical, and even artistic faculties of the human mind."

After such math-enabling brain structures come into play, competitive pressures from equally brainy types could result in the kind of positive feedback that occurs in arms races, rapidly amping up such brainware. At some point, this process might lead to brains capable of registering sophisticated math-related patterns.

These speculations are consistent with findings by cognitive scientists. For instance, studies on the emergence of a basic number sense in infants, as well as brain-imaging studies that have delineated the neural bases of mathematical ability in adults, suggest that we're born with elementary arithmetical abilities. Stanislas Dehaene, a professor at the Collège de France in Paris who has conducted influential research on this topic, theorizes that we possess brain circuits that evolved specifically to represent basic arithmetic knowledge. Such circuits can also support high-level mathematical reflection. He and colleague Marie Amalric showed via brain imaging that high-level math thinking in mathematicians activates a brain network that appears to be largely dedicated to mathematical reasoning.

Importantly, this math-related network is distinct from more recently evolved language centers. That is, humans' mathematical ability appears to be evolutionarily ancient, which suggests that it's the kind of faculty that natural selection fosters early on when conditions are right for human-like intelligence to evolve. This may explain, among other things, why we can often intuit the truth of mathematical statements without being able to articulate exactly why they're true. (Dehaene observes, for instance, that a typical adult can quickly decide that 12 + 15 is not equal to 96 without much introspection on how this cognitive feat is performed.) When this sort of thing happens to mathematicians, they can find themselves obliged to devise proofs for theorems or formulas that they've already sensed must be right. No one has illustrated this phenomenon more vividly than the astounding, self-taught mathematician Srinivasa Ramanujan.

In 1913, Ramanujan, who had lived in poverty during most of his early years in India and had twice failed as a college student, sent off a 10-page manuscript to Cambridge University's G. H. Hardy containing a set of formulas he'd intuited. After perusing the manuscript with growing amazement, Hardy commented that some of the formulas "defeated me completely; I had never seen anything in the least like them before." Although Ramanujan hadn't included proofs, Hardy concluded that most of the strange formulas must have been true because, as he said, "if they were not true, no one would have the imagination to invent them." Hardy was soon telling colleagues that he'd discovered a new Euler in India—like the great Swiss mathematician, Ramanujan had an uncanny ability to sense hidden connections.*

In 1914, Hardy arranged for Ramanujan to come to Cambridge to collaborate with him and his colleague, mathematician J. E. Littlewood. But Ramanujan's ability to dream up jaw-dropping formulas rubbed some math traditionalists the wrong way. They regarded formulas without proofs as probable twaddle, and those who spouted them as something like con artists. Ramanujan also suffered from increasing health

---

*Here's just one of Ramanujan's many provocative formulas: $(1 + 1/2^4) \times (1 + 1/3^4) \times (1 + 1/5^4) \times (1 + 1/7^4) \times (1 + 1/\text{next prime number}^4) \times \cdots = 105/\pi^4$. The infinite product on the left side of this equation is based on successive prime numbers raised to the 4th power. Primes are integers greater than 1 that are evenly divisible only by themselves and 1. Thus, 3 is a prime, but 4 isn't because it's evenly divisible by 2. The first nine primes are 2, 3, 5, 7, 11, 13, 17, 19, and 23. The primes go on forever, which accounts for the ellipsis at the end of the product in Ramanujan's formula. This formula shows a deep connection between π and the prime numbers.

problems after coming to England, including tuberculosis and a severe vitamin deficiency that led to his hospitalization. He died in 1920, soon after his return to India.

One time Hardy came to visit his protégé in a London hospital and experienced an example of Ramanujan's brilliance that has become one of the most famous anecdotes in math history: "I had ridden in taxi cab 1729," Hardy recounted, "and remarked [to Ramanujan] that the number seemed to me rather a dull one, and that I hoped it was not an unfavorable omen. 'No,' he replied, 'it is a very interesting number; it is the smallest number expressible as the sum of two cubes in two different ways.'" Ramanujan was referring to the fact that $1{,}729 = 1^3 + 12^3 = 9^3 + 10^3$, which he'd recorded in one of his notebooks.

Hardy and Ramanujan were both mathematical Platonists, but they personified a schism within the Platonist camp. Hardy was an atheist, and thus to him the independent existence of mathematical truths had nothing to do with divine revelation. Ramanujan believed that his mathematical insights were gifts from the Indian goddess Namagiri, and he famously stated that "an equation has no meaning for me unless it expresses a thought of God."

It seems they could only have agreed to differ on this issue. Indeed, I suspect that the positions that people stake out in the debate about the basic nature of mathematics—like those of Hardy and Ramanujan on secular versus religious Platonism—generally have more to do with axiomatic beliefs than with the close examination of well-established facts. And that's at least partly because the mental processes underlying mathematical intuitions typically aren't directly accessible to the conscious, articulating parts of our minds. Even the most sophisticated brain-imaging studies tell us little about them, and so at this point we can only speculate about why mathematical truths seem possessed of a special kind of inevitability. Still, I like to think that my quasi-Platonist speculations, while only a first step toward an explanation, are on the right track—at least they are consistent with what we know about evolution and the human brain.

BUT WHAT ABOUT INFINITY? We never encounter infinitely large numbers of things, or infinitely small objects, in the real world, and so how might metaphors grounding this crucial mathematical concept arise in sentient beings? Of course, religious-minded Platonists have no problem explaining our notions of infinity: such ideas come to us, according to their basic beliefs, as we channel the thoughts of an infinite-minded God. But I'd prefer an explanation that fits with my secular quasi-Platonism.

It isn't difficult to support the idea that metaphors related to Aristotle's potential infinity bear the stamp of common real-world experiences. For example, the standard Q&A during long car trips with children—"Are we there yet?" "Not yet."—followed after a while by lamentations such as, "It seems like we're never going to get there," suggests a near-universal ability among earth's young sentient beings to conceive things that "never give out in our thought," as Aristotle put it.

But actual infinity seems different. I, for one, can't really get my mind around it. (I'm not referring to the modern mathematical conceptualization of infinity in terms of limits.

Rather, I mean the metaphysical monster—the *Thing*—that visits terrible paradoxes upon us when it's not carefully wrapped up inside the clever evasions of that math framework.) At most, I can summon up an extremely crude idea of actual infinity based on my experiences with large distances (driving across the United States), vast collections of things (sand grains on a beach), and teensy objects (very small specks of dust).

However, the quasi-Platonism I favor doesn't require perfectly exact mappings of independently existing patterns into the mental realm. Even if actual infinity doesn't exist in the real world (and I'm not claiming that it doesn't exist, in some sense; I'm remaining agnostic about that here), independently existing patterns (all those sand grains, etc.) can suggest the idea of it. Thus, I see no reason to exclude actual infinity, and theorems based on it, from the set of concepts that may possess the limited kind of transcendence that I ascribe to mathematics—metaphors for actual infinity would almost certainly occur to the mathematicians of Quasar X9, just as they have on earth.

Ironically, Lakoff and Núñez's theory can be construed as supporting this quasi-Platonist view. They propose that people conceptualize actual infinity by picturing endless processes "as having an end and an ultimate result." That is, we metaphorically blend the idea of completion with process-based potential infinity to conceive actual infinity. They also theorize that mathematicians always use this "Basic Metaphor of Infinity" (BMI) to conceptualize cases of actual infinity that arise in math, such as infinite sets and limits of infinite sums. Their BMI sounds to me like the kind of idea that all sentient beings would probably come up with as they went about both inventing and discovering mathematics.

ALTHOUGH I PART WAYS with Lakoff and Núñez on some issues, I couldn't agree more with the important pedagogical thrust of their book—*Where Mathematics Comes From* makes a strong case for paying more attention to metaphors in math education. Like them, I think we come to understand new things largely by making connections between unfamiliar novelties and familiar mental constructs. Metaphors, broadly defined, help us infer things about novel ideas based on ones that we're familiar with. As Lakoff and Núñez make abundantly clear, mathematics is chock full of such metaphors.

Math's standard terse communication style—all those symbols—makes it possible to compress many conceptual metaphors into a single equation or theorem. That enables great elegance and economy. To students, however, it can seem like a sadistic device dreamed up by a spiteful Numerocracy. Even people who know a lot of mathematics can find it very challenging to understand one of math's intricate concept mashups upon first encountering it. Metaphoric elaboration, as Lakoff and Núñez call it, can be extremely helpful to math learners as they grapple with such high-end mashups.

To show how it's done, they devote a 70-page section of their book to an impressive explication of none other than Euler's formula. But such explanations (including the one offered in these pages) must always be incomplete in an important way, for they can't tell us how Euler sensed the hidden trails that led to the formula. Of course, the proofs he devised provide some clues. But, as suggested by Ramanujan's story, I think mathematical proofs generally represent *ex post facto* glosses on intuitive processes that are mostly inaccessible to our conscious thoughts. A memorable observation by German physicist Heinrich Hertz bears on this point: "One cannot escape the feeling that these mathematical formulas have an independent existence and an intelligence of their own," he wrote, "that they are wiser than we are, wiser even than their discoverers, that we get more out of them than was originally put into them."

I don't see this as an endorsement of Platonic mysticism. It merely highlights the fact that the full meaning of formulas like Euler's is tied to the very deep mystery of how the human mind works. And until that mystery is solved in a detailed way, burning issues about the basic nature of mathematics will probably continue to burn. In any case, no one, as far as I know, has better articulated the basic intuition behind math Platonism (and quasi-Platonism) than Hertz did in this single sentence. He also managed to convey the same sense of wondering delight about the depth, surprise, and beauty of formulas like Euler's that I've experienced, and that I've tried to bring out in this book.

WHILE WRITING ABOUT Euler's formula, I recalled a sculpture I'd seen at the Boston Museum of Fine Arts that struck me as beautifully related to the subject. A kind of epiphany I had while thinking about the sculpture makes a fitting close to the book.

Created by American artist Josiah McElheny, it employs semi-transparent, two-way mirrors to produce reflected rows of bottles, decanters, and other glassware that seem to extend infinitely into the depths of the piece. This visual suggestion of infinity brought to mind the exquisite patterns of the infinite sums that Euler showed are coiled up inside e^{θ}, sin θ, and cos θ.

As I gazed into the piece's visual depths in my mind's eye, however, it occurred to me that the sculpture also represents a concrete metaphor for the sublime profundities that tend to escape our notice as we hurriedly go about our daily routines—until the day, say, that an infant son or daughter is placed in our arms for the first time, or that a beloved person or animal dies, or that one of the most beautiful minds in history reminds us that all the time we're madly rushing around in the unit circles of our days, the infinite, that fantastic figment that can feel so real to us, is quietly lying just beneath the surface.

APPENDIX 1

Euler's Original Derivation

Below, I'll go over Euler's first proof of $e^{i\theta} = \cos\theta + i \sin\theta$. The math is a little more challenging than that in the rest of the book. But most of it will look familiar if you read the trig chapter and are comfortable with basic algebra. (In case of algebraic discomfort, I've included a cheat sheet on pertinent algebra formulas.) And taking on the challenge comes with a major reward: you'll get to look over the shoulder of a genius as he pieces together a great discovery.

First, some math-history mood music.

Euler presented his initial proof of $e^{i\theta} = \cos\theta + i \sin\theta$ in the *Introductio in Analysin Infinitorum* (*Introduction to the Analysis of the Infinite*), a two-volume opus published in 1748. The *Introductio* was basically a high-end pre-calculus text whose goal was to familiarize eighteenth-century math students with the infinite, and the tricky issues it raises, while they were standing on the solid ground of well-known algebraic techniques. Later, they would presumably move on to grappling with infinity in calculus, which at the time was still an evolving branch of math. (Euler later wrote his era's definitive texts on differential and integral calculus.) Among the *Introductio*'s topics were the use of infinitely large and small numbers in calculations; the manipulation of infinite sums; and the expression of trigonometric functions in terms of infinite sums.

Although the *Introductio* was ostensibly a textbook, it was actually more like a very long research paper, and it included a slew of firsts. Examples: it presented the first modern definition of functions; it brought the unit circle to the fore in trigonometry; it pioneered the modern definitions of the sine, cosine, and other trig functions, and established the abbreviations we still use for them, such as sin θ; and it standardized π as the symbol for the famous circle-related number. In a 1950 lecture, historian Carl Boyer called the *Introductio* the most influential mathematics textbook of modern times, comparable to Euclid's *Elements* in importance.

High praise indeed. But French mathematician André Weil arguably topped it in 1979 when he declared that contemporary math students could profit much more from the *Introductio* than from modern alternatives. This probably seemed eccentric to most mathematicians at the time, for it conflicted with the conventional view that the *Introductio*'s derivations are generally based on dubious, eighteenth-century reasoning that, when carelessly applied, can lead to contradictions—such stuff as math nightmares are made on. Post-eighteenth-century mathematicians and historians have sometimes described Euler's conceptual moves as brash, even reckless. Thus, while his results have been celebrated as brilliant advances for over two centuries, many of his derivations, particularly ones related to infinity, have long been regarded as little more than quaint relics. Indeed, writers on math have often marveled over the fact that he almost always reached what are now regarded as sound conclusions—he seemed to have had an uncanny knack for doing the right thing despite his purportedly dubious reasoning. In effect, while belittling Euler's methods, they've bolstered the idea that he possessed almost preternatural intuition.

But Weil's view has seemed less eccentric in recent years. One reason is that revisionist mathematicians have shown that Euler's inferential moves when proving infinity-related theorems were actually quite similar to ones that are now made in non-standard analysis, an indisputably rigorous branch of mathematics that emerged in the 1960s. (Non-standard analysis is based on "hyperreal" numbers, which are basically the infinitely large and small numbers of Euler's era couched in rigorous, modern terms.) Their research suggests that most of Euler's conceptual maneuvering, with only minor adjustments, can be seen as perfectly valid by today's standards.

Further, some math educators have argued that Euler's conceptual moves are better aligned with students' natural intuitions, and thus easier to grasp, than their standard modern counterparts. (This view isn't new, by the way, but its appeal has grown as teachers have cast about for better ways to introduce calculus.) Today, math texts that make reference to infinity often read like legal documents filled with abstruse logic and complex qualifying clauses. The complexities help ward off inconsistencies that were implied by the less-than-rigorous concepts of Euler's day—these older ideas were basically trashed by nineteenth-century mathematicians as they sought to eliminate shadows of doubt that had loomed over math's treatment of the infinite. But the greater rigor had a cost: it distanced a lot of math from intuitively appealing ideas that had long informed mathematical thought. For instance, picturing functions as curves that a moving hand might draw came to be seen as relying on ill-defined geometrical intuitions that could all too easily lead to serious weirdness, or even horrible cracks, in math's very foundations.*

Euler's masterpieces are also noteworthy because of their lucidity. Indeed, he's almost unique in mathematics for "taking pains" to carefully present his reasoning, observed twentieth-century Hungarian mathematician George Pólya. Largely because of that, Pólya added, Euler's works possess a "distinctive charm"—a quality that's not superabundant in the technical math lit. Mathematician William Dunham has similarly noted that Euler's expositions are "fresh and enthusiastic, in contrast to the modern tendency of obscuring a scholar's passion behind the facade of detached technical prose."

TO UNDERSTAND HOW Euler derived $e^{i\theta} = \cos \theta + i \sin \theta$, you need to know a little about the infinitely small numbers—

*Here's an example of the kind of weirdness that geometrical intuitions can lead to: Picture two, same-sized adjacent circles touching at a single point. It's intuitively natural to think of them as "kissing"—you might call it "Circles in Love" if you were a minimalist-art aficionado. But now consider what this kiss would actually be in human terms: two people whose lip skin melts together when they kiss, so that they're literally joined at the lips. This follows from the fact that the point where the circles touch is an intersection point like one shared by two straight lines that cross—it lies on both circles. George Lakoff and Rafael E. Núñez put this wonderfully creepy insight on the record in their book *Where Mathematics Comes From*. It isn't mathematically troublesome, but it memorably illustrates how seemingly simple geometric ideas sometimes imply strange things.

infinitesimals—that he routinely introduced in calculations. Infinitesimals were popularized in mathematics by Leibniz when he developed his version of calculus in the 1670s. They were loosely defined as numbers that were so very close to zero that they could be treated as zeroes—or not—depending on the circumstances. Importantly, when wearing their non-zero hats, they could serve as divisors, or, equivalently, as denominators of fractions. In contrast, true zeroes can't do that—dividing by zero isn't allowed in math, and fractions of the form $x/0$ are undefined.

These vaguely defined numberlets were crucial in the development of calculus. And following in Leibniz's footsteps, Euler and his contemporaries freely introduced them when computing instantaneous rates of change (which involve zero time increments, because that's what the term "instant" means). That enabled them to avoid dividing by zero in such calculations—instead, they divided by infinitesimals. But when the convenient specks were no longer needed, they would be eliminated as a simplifying move, just as if they really were equal to zero after all. This maneuver resembled the perfectly valid move of replacing $x + 0$ with x, and it allowed mathematicians to write things like $x + dx = x$, where dx designates a non-zero (wink, wink) infinitesimal. Such strategic "neglect" of infinitesimals was justified by the argument that they were too small to matter when added or subtracted from what might be called normal-sized (finite) numbers in calculations.

Plainly, the infinitesimals of Euler's day were metaphysically fishy. But like a zillion magical microbes in harness, they helped propel a powerful engine of discovery. In fact, "great creations" in mathematics were "more numerous [during the eigtheenth century] than during any other century," according to eminent math historian Morris Kline.

Kline added, however, that Euler and his peers were so "intoxicated" with their successes that they were often "indifferent to the missing rigor" of their mathematics. The most glaring problem was that their clever calculating tricks had skirted, rather than solved, deep problems involving the infinite that had bedeviled mathematics and philosophy since the time of the ancient Greeks. By 1900, fully sobered-up mathematicians had replaced the loose, intuitive ideas of Euler's time with precision-engineered definitions of limits and other concepts that referenced only finite quantities, effectively shoving the threatening specters out of sight.* (Although not necessarily out of mind—the infinite will probably always be a very provocative topic.)

NOW FOR THE MATH. Let's begin with a set of rules governing the use of infinitely small and large numbers in Euler's

*For example, here's the definition of the limit of the infinite sequence of fractions, 1, 1/2, 1/3,... (i.e., $1/n$, where $n = 1, 2, 3,...$): The sequence $1/n$ for $n = 1, 2, 3,...$ has the limit L if, for any positive number ε, there exists a positive integer m such that the absolute value of $L - 1/n$ is less than ε for all n greater than or equal to m. (For this sequence, by the way, $L = 0$.) Complicated, yes. Infinity-infested, well, not exactly. The infinite—that is, the infinitely small—is hidden behind the phrase "for any positive ε," which implies that we can make $1/n$ as close to L as we like by choosing n to be ever larger. Importantly, this "epsilontic" definition (so-named because ε, the Greek letter epsilon, is frequently used in math to designate a small, finite number) omits the somewhat vague notion of motion (conveyed by terms such as "approaching" or "tending toward"), which was implicit in earlier definitions of limits. The now-standard, industrial-strength definition of limits was formulated by German mathematician Karl Weierstrass.

reckonings. Just give them a quick once-over at this point; you can revisit them as needed when they're invoked in the calculations below.

(1) Multiplying an infinitely small number times a finite one yields another infinitesimal. This is analogous to the rule of arithmetic specifying that multiplying a number times zero equals zero. Thus, if y and z designate infinitesimals, and x is a finite number, you can write $y \times x = z$, or $yx = z$.

(2) Dividing a finite number by an infinitely large one yields an infinitesimal. This logic is akin to dividing a cake into an infinitely large number of pieces for the attendees at a very, very large birthday party, resulting in each person getting a vanishingly small piece. Mathematically, it would be $x/n = z$, where n is infinitely large, x is a finite number, and z is an infinitesimal.

(3) Dividing a positive finite number by a positive infinitesimal is equal to an infinitely large number. This is analogous to dividing a positive number by a very small positive fraction to get a very big number. For instance, 1 divided by a millionth (which is equivalent to the number of one-millionths there are in the number 1), equals a million, or, in short, $1/(1/1{,}000{,}000) = 1{,}000{,}000$. Thus, if x were a finite number, z were an infinitesimal, and n were an infinitely large number, you could write $x/z = n$.

(4) Multiplying an infinitesimal times an infinitely large number yields a finite number. There's no analogue to this rule in regular arithmetic. However, it accords with the intuitive idea that when the infinitely large is pitted against the infinitely small, the two basically cancel each other out in a titanic clash, and after the dust clears a finite number remains.

Thus, if z designates a positive infinitesimal, n stands for an infinitely large number, and x represents a finite number, you could write $z \times n = x$, or $zn = x$.

(5) Recall that $\cos 0 = 1$ and $\sin 0 = 0$. Because the difference between an infinitesimal, call it z, and 0 is infinitely small, it would make sense—at least when no post-eighteenth-century mathematician is looking—to sneak z in for 0 in these trig facts to get $\cos z = 1$ and $\sin z = z$. This move, which Euler made in the derivation at hand, invokes one side of the two-faced nature of infinitesimals—in effect, z is treated like zero here. George Berkeley, the philosopher who argued that infinitesimals were metaphysically goofy, is spinning in his grave about now. But Euler is remaining totally cool with the whole thing during his infinitely long rest.

HERE'S THE PROMISED algebra cheat sheet. These rules show how to manipulate terms (designated by a, b, c, and d) that represent specific numbers, individual variables, or elaborate expressions involving numbers and variables.

- $a^0 = 1$, $a^1 = a$, $a^2 = a \times a$, $a^3 = a \times a \times a$, etc.
- $(a^m)^n = a^{m \times n}$
- If $a = b$, then $a^n = b^n$
- If $a/b = c$, then $a/c = b$.
- $a/b \times b = a$
- If $a = b$ and $c = d$, then $a + c = b + d$ (Which, in effect, means that equations can be added together.)
- $(a + b) \times (c + d) = (a \times c) + (a \times d) + (b \times c) + (b \times d)$ (This is known as the FOIL rule, for it entails expanding the

left side of the equation to get the right side by successively adding the products of: the First components of each term on the left (*a* and *b*), the Outer components (*a* and *d*), the Inner components (*b* and *c*), and the Last components (*b* and *d*).)

- $(a + b)/2 = a/2 + b/2$.

I'VE DIVIDED EULER'S derivation into seven steps and labeled certain equations (e.g., A.1 and A.2 in the first step) so they can be easily referenced later. I've also updated Euler's notation—I write, for instance, x^2 instead of xx as he did for typesetters' convenience. With one exception, noted in Step 6, the reasoning presented here closely tracks Euler's.

Step 1: Our first move is to show how de Moivre's formula, which was mentioned in passing in Chapter 9, can be extracted from a couple of basic trigonometry equations called the angle addition identities. I've opted to skip derivations of these identities; they follow from the triangle-based trig I showed you in Chapter 7, and if you want to see the proofs, I recommend the Khan Academy's excellent versions at www.khanacademy.org. (Just search its website for "proof of angle addition identities.")

The identities are:

$$\sin(a + b) = \sin a \cos b + \sin b \cos a$$

and

$$\cos(a + b) = \cos a \cos b - \sin a \sin b.$$

In case you're wondering, the terms on the right sides of these equations represent products. For instance, $\sin a \sin b = \sin a \times \sin b$.

To generate de Moivre's formula, Euler used the identities to show that expressions of the form $(\cos\theta + i\sin\theta)^n$ are equal to expressions of the form $\cos(n\theta) + i\sin(n\theta)$, where $n = 1, 2, 3, \ldots.$

This is trivial to prove for $n = 1$, because

$(\cos\theta + i\sin\theta)^1 = \cos\theta + i\sin\theta = \cos(1\times\theta) + i\sin(1\times\theta)$ (Since $a^1 = a$, and $1\times\theta = \theta$.)

For $n = 2$, we have

$(\cos\theta + i\sin\theta)^2 = (\cos\theta + i\sin\theta)(\cos\theta + i\sin\theta)$

$= \cos^2\theta + i\sin\theta\cos\theta + i\sin\theta\cos\theta + i^2\sin^2\theta$ [Using FOIL, and the fact that $\cos^2\theta$ means $\cos\theta$ times $\cos\theta$ —likewise for $\sin^2\theta$.]

$= (\cos^2\theta - \sin^2\theta) + [i\times(2\sin\theta\cos\theta)]$ [By using the fact that $i^2 = -1$, combining the first and fourth terms after converting i^2 to -1, and by adding the two identical middle terms.]

$= \cos 2\theta + i\sin 2\theta$,

because of the angle addition identities given above. For instance, when both a and b are set equal to θ in the second identity, it implies that $\cos 2\theta = \cos(\theta+\theta) = \cos\theta\cos\theta - \sin\theta\sin\theta = \cos^2\theta - \sin^2\theta$, which justifies replacing $\cos^2\theta - \sin^2\theta$ with $\cos 2\theta$.

For $n = 3$, we have

$(\cos\theta + i\sin\theta)^3 = (\cos\theta + i\sin\theta)^2(\cos\theta + i\sin\theta)$ [By the definition of exponents.]
$= (\cos 2\theta + i\sin 2\theta)(\cos\theta + i\sin\theta)$ [By plugging in the result above for $n = 2$.]
$= \cos 2\theta\cos\theta + i\cos 2\theta\sin\theta + i\sin 2\theta\cos\theta - \sin 2\theta\sin\theta$ [By FOIL and $i^2 = -1$.]
$= (\cos 2\theta\cos\theta - \sin 2\theta\sin\theta) + [i \times (\cos 2\theta\sin\theta + \sin 2\theta\cos\theta)]$ [By rearranging terms.]
$= \cos 3\theta + i\sin 3\theta$ [By using the trig identities with $a = 2\theta$ and $b = \theta$].

For $n = 4$, we repeat this procedure, duly plugging in the result we got for $n = 3$ and invoking the trig identities with a and b set equal to 3θ and θ, respectively, to obtain

$$(\cos\theta + i\sin\theta)^4 = \cos 4\theta + i\sin 4\theta.$$

At this point, I hope you can see that repeating the procedure for $n = 5, 6$, and so on will yield similar results. Expressing this conclusion for any positive integer n gives us de Moivre's formula:

$$\text{(A.1)} \quad (\cos\theta + i\sin\theta)^n = \cos(n\theta) + i\sin(n\theta).$$

Tweaking A.1 yields a similar equation:

$$\text{(A.2)} \quad (\cos\theta - i\sin\theta)^n = \cos(n\theta) - i\sin(n\theta).$$

The tweak simply involves replacing θ by $-\theta$ in A.1, and then applying these trig facts: $\sin(-\theta) = -\sin\theta$, and $\cos(-\theta) = \cos\theta$. You can verify the trig facts, if desired, by reviewing Chapter 7's unit-circle-based definition of the trig functions and the meaning of negative angles. Or visit the Khan Academy and search for "Sine & cosine identities: symmetry."

Step 2: Reversing equations A.1 and A.2 and adding them together, we get

$$\cos(n\theta) + i\sin(n\theta) + \cos(n\theta) - i\sin(n\theta) = (\cos\theta + i\sin\theta)^n + (\cos\theta - i\sin\theta)^n.$$

Rearranging this equation's left-side terms makes it $\cos(n\theta) + \cos(n\theta) + i\sin(n\theta) - i\sin(n\theta)$, or simply $2\cos(n\theta)$. (Note that $i\sin(n\theta) - i\sin(n\theta)$ equals 0—the two terms cancel each other out.) Thus, the equation becomes

$$2\cos(n\theta) = (\cos\theta + i\sin\theta)^n + (\cos\theta - i\sin\theta)^n,$$

and, by dividing each side by 2, we obtain

$$\text{(A.3)} \qquad \cos(n\theta) = [(\cos\theta + i\sin\theta)^n + (\cos\theta - i\sin\theta)^n]/2.$$

At this point, Euler brought infinity into play. He assumed that the n shown in equation A.3 is infinitely large. If a finite number, which we'll represent by the variable v, is divided by the infinitely large n, the result is an infinitesimal number, which we'll call z. In short, $v/n = z$ (in accordance with Rule

2 above). By basic algebra, $v/n = z$ implies that $v = zn = nz$. (Which accords with Rule 4.)

Note well: n, z, and v will be brought into play multiple times below.

Now, by applying Rule 5 to z, we have $\cos z = 1$. Using the "sine part" of the same rule along with $z = v/n$ (from the previous paragraph), we get, as Euler did, $\sin z = z = v/n$.

Hang on, we're making real progress. We're now close to turning the sines and cosines on the right side of equation A.3 into expressions very much like $(1 + 1/n)^n$ with n assumed to be infinitely large. This expression is the number e, which we need to bring into the picture in order to produce the equation we're after.

Here's a quick review of the numbers that are now available for action: we've assumed that n is an infinitely large number, z is an infinitely small one, $nz = v$ (where v is finite), $z = v/n$, $\cos z = 1$, and $\sin z = v/n$.

The next maneuver is to plug in an infinitely small number for θ in A.3, namely z. After we plug in z for θ on A.3's right side, it becomes $[(\cos z + i \sin z)^n + (\cos z - i \sin z)^n]/2$. Now comes a key move that Euler cleverly set up by bringing the infinite into play: because $\cos z = 1$ and $\sin z = v/n$, we can replace $\cos z$ with 1, and $\sin z$ with v/n, to turn A.3's right side into $[(1 + iv/n)^n + (1 - iv/n)^n]$. (Note the two e-like thingies we've now conjured up.)

Meanwhile, when we plug in z for θ on A.3's left side, it becomes $\cos (nz)$, and since $nz = v$ the left side can be rewritten as $\cos v$.

Installing these new, improved left and right sides in A.3 gives us

(A.4) $\cos v = [(1 + iv/n)^n + (1 - iv/n)^n]/2.$

Step 4: This step is virtually identical to Step 3, except that its first move is to subtract equation A.2 from A.1 instead of adding the two. That yields

$$2i \sin (n\theta) = (\cos \theta + i \sin \theta)^n - (\cos \theta - i \sin \theta)^n,$$

which, by duplicating the infinity-based logic of Step 3, leads to

(A.5) $i \sin v = [(1 + iv/n)^n - (1 - iv/n)^n]/2.$

Step 5: Now let's clear away some clutter by temporarily replacing the expressions $(1 + iv/n)^n$ and $(1 - iv/n)^n$ in equations A.4 and A.5 with the variables r and t, respectively. That turns them into

$$\cos v = (r + t)/2$$

and

$$i \sin v = (r - t)/2.$$

Adding these equations together gives us

(A.6) $\cos v + i \sin v = (r + t)/2 + (r - t)/2.$

By basic algebra, $(r + t)/2 = r/2 + t/2$, and $(r - t)/2 = r/2 - t/2$, which allows us to rewrite A.6 as

$\cos v + i \sin v = r/2 + t/2 + r/2 - t/2.$

Finally, notice that the second and fourth terms of the right side of this last equation cancel each other out, and the first and third add up to r. This means that the right side algebraically melts down to r alone, which, recall, is equal to $(1 + iv/n)^n$. And that permits rewriting A.6 again as

(A.7) $\cos v + i \sin v = (1 + iv/n)^n.$

Step 6: Now all we need to do is show that the right side of A.7 is equivalent to e^{iv} and we'll be done. Continuing in Euler's footsteps, we'll begin by assuming that a is a finite number greater than 1 and z is an infinitely small number. From algebra, we know that $a^0 = 1$, $a^1 = a$, and, in general, increasing a's exponent will produce larger and larger numbers. (For instance, if $a = 3$, then $a^0 = 1$, $a^1 = 3$, $a^2 = 9$, and so on.) Thus, since z, an infinitesimal, is assumed to be only very slightly greater than 0, it stands to reason that a^z will be only very slightly greater than a^0. Indeed, Euler reasoned that a^z is only infinitesimally greater than a^0. Expressing this as an equation, we have $a^z = a^0 + w$, where w is infinitely small. And since $a^0 = 1$, we can rewrite this equation as $a^z = 1 + w$.

Dividing a number m by the number p to get k can be written as $m/p = k$. Similarly, for the infinitesimals w and z, we can write $w/z = k$ for some number k. Multiplying both sides of this equation by z and applying basic algebra to the result gives us $w = kz$. This means that we can replace w with kz in the equation above ($a^z = 1 + w$) to get

(A.8) $a^z = 1 + kz.$

Importantly, this last equation implies that a and k are tied to each other in the same way that y and x are tied by the equation $y = 1 + 2x$—in the latter, if x is assigned a specific value, say 2, then y must take on a corresponding specific value, in this case 5. This important a-k linkage will be revisited momentarily.

Now glance back at Rule 2 above. It implies that the infinitesimal z is equal to some number v/n, where v stands for a finite number (the variable x was used to represent a finite number above, but v will work just as well), and n stands for an infinitely large number. This logic expressed as an equation is $z = v/n$, which allows us to replace z with v/n in A.8 to get

$$a^{v/n} = (1 + kv/n).$$

Raising both sides of this equation to the nth power yields $(a^{v/n})^n$ on the left side, which, by applying a bit more algebra, we reduce to a^v. Meanwhile, raising the equation's right side to the nth power turns it into $(1 + kv/n)^n$. Thus, we now have:

(A.9) $a^v = (1 + kv/n)^n.$

Based on the a-k linkage mentioned above, we know that if we set k equal to a specific number, then a must take on a corresponding specific value. Thus, if we set k equal to 1, a will take on the value of some corresponding constant. To identify this mystery constant, let's examine the equation after k is set equal to 1:

$a^v = (1 + v/n)^n.$

Because v represents an unspecified finite number, it can freely range over different numbers without falsifying the equation. That permits us to set v equal to 1 if we want (and we do), turning the equation into

$a = (1 + 1/n)^n.$

Now we can see what a must be when $k = 1$. Based on the definition of e from Chapter 2 (and the fact that n was assumed above to be infinitely large), this last equation implies that $a = e$. Therefore, we can rewrite A.9 with $k = 1$ and $a = e$ to get

(A.10) $e^v = (1 + v/n)^n.$

(Euler used more elaborate reasoning than shown here to get to A.10. But the trail he followed also led to the conclusion that $a = e$ when $k = 1$.)

Last step: Recall from Chapter 9 how Euler boldly substituted the imaginary-number variable $i\theta$ for the real-number variable θ in an equation he knew was true for real numbers. Let's do the same thing in equation A.10 by plugging in iv for the real-number variable v to get

$e^{iv} = (1 + iv/n)^n,$

which, with A.7, implies that e^{iv} and $\cos v + i \sin v$ equal the same thing, namely $(1 + iv/n)^n$. Thus, e^{iv} and the expression cos

$v + i \sin v$ are themselves equal, or, as Euler rather excitedly put it when he arrived at his famous formula in the *Introductio*, "truly there will be" (*erit vero*)

$$e^{iv} = \cos v + i \sin v,$$

or, more familiarly, $e^{i\theta} = \cos \theta + i \sin \theta$, using θ as the variable.

APPENDIX 2

Why i^i Is Real

The one-unit imaginary number with itself as an exponent, i^i, seems strictly unreal. But its looks are quite deceiving. Here's how Euler's formula can be used to show that it's a real number:

First, notice that when equal numbers expressed in different ways are raised to the same power, the resulting numbers are also equal. Example: Since $4/2 = 2$, we know that $(4/2)^2 = 2^2$. Next, recall from Chapter 11 that $e^{i\pi/2} = i$. (It can be derived by plugging $\pi/2$ into $e^{i\theta} = \cos\theta + i \sin\theta$.) If we raise both of these equal numbers to the *i*th power, we should get two numbers that are also equal. That is, $(e^{i\pi/2})^i = i^i$. This equation means that evaluating $(e^{i\pi/2})^i$ will reveal the numerical identity of i^i.

Now, to evaluate $(e^{i\pi/2})^i$, consider a similar expression, 2^2 raised to the third power, or $(2^2)^3$, which can be written $(2 \times 2)^3$, or $(2 \times 2) \times (2 \times 2) \times (2 \times 2)$, which equals 2^6, or $2^{2 \times 3}$. This example illustrates a general rule that can be stated with variables as $(x^a)^b = x^{a \times b}$. Applying the rule to $(e^{i\pi/2})^i$ gives us $(e^{i\pi/2})^i = e^{i\pi/2 \times i} = e^{i \times i \times \pi/2}$ (by rearranging multiplied terms in the exponent) $= e^{-\pi/2}$ (since, $i \times i = i^2 = -1$). Thus, $i^i = e^{-\pi/2}$, and although $e^{-\pi/2}$ has a funny-looking negative exponent, it is a real number—it's actually about 0.208. (In fact, i^i is equal to an infinity of real numbers related to this decimal, which is termed the principal value of i^i, but that's another story.) When Euler discovered that i^i is real, he exclaimed in a letter to a friend that this "seems extraordinary to me"—part of his genius, as well as of his charm, was an inexhaustible capacity to be surprised and delighted by his discoveries.

ACKNOWLEDGMENTS

Several non-mathematicians gave me invaluable feedback on this book as it took shape: Bill Bulkeley, Nancy Malle, Alicia Russell, and John Russell. Mathematicians William Dunham, an authority on Euler and his work, and Underwood Dudley, who formerly edited the *College Mathematics Journal*, were immensely helpful and generous with their time, acting both as my consultants on math issues and as eagle-eyed editors on matters of notation, wording, and style. (But of course I'm solely responsible for whatever might be wrong with the book.) I'm also indebted to the creators and sustainers of the Euler Archive, a rich online library that saved me a lot of research time; to Sal Khan, originator of the Khan Academy, which offers superb explanations that I looked to as models; and to John J. O'Connor and Edmund F. Robertson, creators of the wonderfully comprehensive online MacTutor History of Mathematics, which I heavily relied on.

Fellow math enthusiast Christoph Drösser provided needed encouragement early on as I considered the daunting challenge of writing about a deep mathematical discovery in a way that might appeal to math-averse people. He also read a draft and helped improve it. My agent, Lisa Adams, was also very encouraging. Basic Books' editor T.J. Kelleher, who has a rare gift for both getting words right and grasping the essence of technical topics, played a *sine-qua-non* role in making the book happen. I'm also beholden to the talented team that transformed the manuscript into a book, including Melissa Raymond, Michelle Welsh-Horst, Marco Pavia, Brent Wilcox, and Samantha Manaktola.

My wife, Alicia Russell, provided key infusions of equanimity as I wrestled with, and often got thrown by, my muse. Our kids aided me too. Claire, who was in sixth grade as I wrote the book, helped convince me that a person knowing no more math than sixth graders do can understand basic trigonometry and other concepts needed to make sense of Euler's formula. And Quentin, a young artist who now works for computer-game and film companies, helped me figure out how to explain math to people who tend to shy away from it by somehow putting up with my gung-ho math coaching all the way from first-grade arithmetic through high-school calculus.

GLOSSARY

associative law of addition: Typically written $a + (b + c) = (a + b) + c$, this rule means that it doesn't matter how you group added numbers, or which two you add first.

associative law of multiplication: Written $(a \times b) \times c = a \times (b \times c)$, it means that it doesn't matter how you group multiplied numbers, or which two you multiply first.

commutative law of addition: Written $a + b = b + a$, it means switching the order of two numbers when adding them doesn't change the result.

commutative law of multiplication: Written $a \times b = b \times a$, it means that switching the order of two numbers when multiplying them doesn't change the result.

complex plane: A two-dimensional space containing two number lines, or axes, that intersect at right angles—an x axis for real numbers, and an i axis for imaginary numbers.

complex number: Two-part, hybrid numbers that are usually written in the form $a + bi$, where a is a real number and bi, or b times i, is an imaginary number. Each complex number is associated with a point on the complex plane.

constant: A particular number, such as e, i, π, 1, or 0, which are the five constants in Euler's formula.

cosine function: Written cos θ, it's a function that effectively takes as input the size of an angle of a right triangle and that outputs the ratio between the length of the side adjacent to the angle and the length of the triangle's hypotenuse. Like the sine function, it can also be defined in terms of the coordinates of points on the unit circle.

cubic equation: An equation that includes a variable such as x raised to the third power (that is, having an exponent of three) but no x's raised to higher powers. Example: $x^3 + 2x^2 - 5x + 8 = 0$.

distributive law: An arithmetic rule often written $a \times (b + c) = (a \times b) + (a \times c)$. It means that when you multiply a sum by a number, you get the same result that you would get if you separately multiplied the number times each of the summed numbers and then added the products together.

***e*:** A constant, defined as the number that the expression $(1 + 1/n)^n$ approaches as ever larger integers are plugged in for n. Called Euler's number, it often crops up in math, sometimes quite unexpectedly. It is an irrational number as well as a transcendental one.

exponent: In basic math, an exponent is defined as a positive integer written as a superscript next to a constant or variable that designates how many times to multiply the constant or variable times itself. For instance, 10^2, which is spoken "ten squared" or "ten raised to the second power," means 10 times 10, or 100, and 10^3 means 10 times 10 times 10, or 1,000. Mathematicians have repeatedly expanded this definition to allow zero, negative integers, non-integer real numbers (both rational and irrational), imaginary numbers, and complex numbers as exponents.

factorial: Symbolized by ! written adjacent to an integer, the factorial operator means "multiply together all the positive integers up to and including the specified integer." Thus, 3!, which is spoken "three factorial," is shorthand for $1 \times 2 \times 3$, or 6. Both 0! and 1! are defined as 1.

function: As used in this book, the term function means an expression with variables, such as $x + 5$. Functions resemble computer programs that convert input numbers into output numbers in specified ways. They're designated with equations such as $f(x) = x + 5$, where $f(x)$ means "a function whose variable is x."

hypotenuse: The side opposite the 90-degree angle of a right triangle.

***i*:** Defined as the square root of –1, i is the one-unit imaginary number on which all the imaginary numbers are based.

imaginary number: A number of the form $a \times i$, where a is a real number, and i stands for the square root of –1. Each imaginary number is the counterpart of a real number. For instance, i, the unit imaginary number, is the imaginary counterpart of 1, and $-i$ is the imaginary counterpart of –1. The imaginary number π times i is e's exponent in Euler's formula.

irrational number: A number that can't be expressed as a fraction. An irrational number's decimal representation includes an infinite number of patternless digits to the right of the decimal point. Both π and e are irrational numbers.

***n*th root:** The nth root of a number is another number that when multiplied times itself n times is equal to the original number. For instance, 2 is a 4th root of 16, since $2 \times 2 \times 2 \times 2 = 16$.

oscillation: Movement back and forth at a constant speed. Cyclic phenomena such as sound and radio waves, and alternating current, involve oscillation.

origin: The point at which the x and y axes meet in the xy plane, and at which the x and i axes meet in the complex plane. Its coordinate pair is (0,0) in the xy plane. The complex plane's origin is associated with the complex number $0 + 0i$.

parallelogram: A four-sided polygon whose opposite sides are parallel.

pi: Designated by the Greek letter π, pi is a constant that's equal to the circumference of any circle divided by its diameter. An irrational number, it's approximately 3.14159.

polygon: A many-sided figure. Regular polygons have equal-length sides and equal-sized internal angles, such as stop-sign octagons.

protractor: A tool for measuring angles, typically consisting of a half circle made of transparent plastic, with angles marked off on the outer edge from 0 to 180 degrees.

Pythagorean theorem: It states that the sum of the squared lengths of the two shorter sides of any right triangle is equal to the square of the hypotenuse's length. It is typically expressed with an equation like $x^2 + y^2 = z^2$, where x, y, and z represent side lengths of a right triangle.

radian: The angle swept out from the center of a circle by an arc along the circle that's equal in length to its radius. This implies that 2π radians = 360°, π radians = 180°, and $\pi/2$ radians = 90°.

radius: Half of a circle's diameter, pictured as a line segment between the circle and its center.

ratio: Typically written in the format 2 to 3, or 2:3, ratios express the same numerical relationships that fractions do. For instance, if a recipe specified a sugar-to-flour ratio of 1 to 3, you could say that it calls for a mixture consisting of 1/4 sugar and 3/4 flour.

real number: A number that lies along the familiar number line. The reals include positive and negative integers, zero, rational numbers (fractions), irrational numbers, and transcendental numbers. The rationals include all the integers; the irrationals include the transcendentals.

right triangle: A triangle with a 90-degree angle and two smaller angles.

sine function: Written sin θ, it's a function that effectively takes as input the size of an angle of a right triangle and that outputs the ratio between the length of the side opposite to the angle and the length of the triangle's hypotenuse. Like the cosine function, it can also be defined in terms of coordinate pairs of points on the unit circle.

square root: The square root of a number x is a number that when multiplied times itself equals x. For example, both 2 and –2 are square roots of 4.

transcendental number: A number that cannot be the solution of any polynomial equation with integer constants multiplied times the variables. An example of a polynomial equation is $x^2 - 2x - 35 = 0$. The fact that 7 solves this equation rules out that number as a transcendental. Both π and e were demonstrated in the nineteenth century to be transcendental.

unit circle: A one-unit-radius circle whose center is the origin of the xy plane, or the origin of the complex plane. Radius-like line segments can be pictured in the unit circle sweeping out angles.

variable: Usually designated by letters such as x, variables are proxies for numbers that haven't been pinned down. When they appear in equations such as $x - 2 = 4$, they represent "unknowns" that can be determined by finding numbers that, when substituted for the variables, make the equations true.

vector: An arrow-like visual representation of complex numbers on the two-dimensional complex plane. Vectors can also be used as visual proxies of coordinate pairs on the xy plane.

Vectors are used in physics to represent things such as the speed and direction of moving objects.

***x* and *y* coordinates:** Pairs of numbers written in parentheses that are associated with points on the *xy* plane. The first of the two coordinates is measured along the *x* axis, specifying a point's distance from the y axis. The second coordinate is measured along the *y* axis and specifies the point's distance from the *x* axis.

***xy* plane:** A two-dimensional surface featuring a horizontal real-number line called the *x* axis, and a vertical real-number line called the *y* axis. The *xy* plane enables the mapping of arithmetic and algebraic concepts onto geometric counterparts, such as the representation of coordinate pairs of numbers as points on the plane.

Zeno's paradox: Zeno, an ancient Greek philosopher, proposed several thought experiments that led to preposterous conclusions. The "race course paradox," one of his most famous, suggests that a runner can never complete a race because he must first get half of the way to the finish line, then half of the remaining distance after that, and then half of the remainder after that, and so on—an infinite number of steps that was seemingly impossible to complete.

BIBLIOGRAPHY

Allain, Rhett. "Modeling the Head of a Beer." *Wired*, January 25, 2009. https://www.wired.com/2009/01/modeling-the-head-of-a-beer/.

Amalric, Marie, and Stanislas Dehaene. "Origins of the Brain Networks for Advanced Mathematics in Expert Mathematicians." *Proceedings of the National Academy of Sciences*, May 3, 2016, 4909-4917.

Apostol, Tom M. *Calculus, Volume I* and *Volume II*. Waltham, Mass.: Blaisdell Publishing Co., 1969.

Archibald, Raymond Clare. *Benjamin Peirce, 1809-1880: Biographical Sketch and Bibliography*. Oberlin, Ohio: The Mathematical Association of America, 1925. https://archive.org/details/benjaminpeirce1800arch.

Assad, Arjang A. "Leonard Euler: A Brief Appreciation." *Networks*, January 9, 2007. http://onlinelibrary.wiley.com/doi/10.1002/net.20158/abstract.

Bair, Jacques, Piotr Blaszczyk, Robert Ely, Valerie Henry, Vladimir Kanovei, Karin U. Katz, Mikhail G. Katz, Semen S. Kutateladze, Thomas McGaffey, Patrick Reeder, David M. Schaps, David Sherry, and Steven Shnider. "Interpreting the Infinitesimal Mathematics of Leibniz and Euler," *Journal for General Philosophy of Science*, 2016. https://arxiv.org/abs/1605.00455.

Baker, Nicholson. "Wrong Answer: The Case against Algebra II." *Harper's*, September 2013. http://harpers.org/archive/2013/09/wrong-answer/.

Ball, W. W. Rouse. *A Short Account of the History of Mathematics*. New York: Dover Publications Inc., 1908.

Bell, E. T. *Men of Mathematics: The Lives and Achievements of the Great Mathematicians from Zeno to Poincaré*. New York: Touchstone Books, 1986.

Bellos, Alex. *The Grapes of Math: How Life Reflects Numbers and Numbers Reflect Life*. New York: Simon & Schuster, 2014.

Benjamin, Arthur. *The Magic of Math: Solving for x and Figuring Out Why*. New York: Basic Books, 2015.

Blatner, David. *The Joy of Pi*. New York: Walker & Co., 1997.

Bouwsma, O. K. *Philosophical Essays*. Lincoln, Neb.: University of Nebraska Press, 1965.

Boyer, Carl B. "The Foremost Textbook of Modern Times." Mac Tutor History of Mathematics archive, 1950. http://www-groups.dcs.st-and.ac.uk/~history/Extras/Boyer_Foremost_Text.html.

Bradley, Michael J. *Modern Mathematics: 1900-1950*. New York: Chelsea House Publishers, 2006.

Bradley, Robert E., Lawrence A. D'Antonio, and C. Edward Sandifer, Editors. *Euler at 300: An Appreciation*. Washington, D.C.: The Mathematical Association of America, 2007.

Branner, Bodil, and Nils Voje Johansen. "Caspar Wessel (1745-1818) Surveyor and Mathematician." In *On the Analytical Representation of Direction: An Attempt Applied Chiefly to Solving Plane and Spherical Polygons* by Caspar Wessel, trans. from the Danish by Flemming Damhus, edited by Bodil Branner and Jesper Lutzen, 9-61. Copenhagen: Royal Danish Academy of Science and Letters, 1999.

Burris, Stan. "Gauss and Non-Euclidean Geometry," 2009. https://www.math.uwaterloo.ca/~snburris/htdocs/geometry.pdf.

Cajori, Florian. "Carl Friedrich Gauss and his Children." Science, May 19, 1899, 697-704, https://www.jstor.org/stable/1626244?seq=1#page_scan_tab_contents.

Calinger, Ronald S. *Leonhard Euler: Mathematical Genius in the Enlightenment*. Princeton: Princeton University Press, 2015.

Clegg, Brian. *A Brief History of Infinity: The Quest to Think the Unthinkable*. London: Constable & Robinson Ltd., 2003.

Debnath, Lokenath. *The Legacy of Leonhard Euler: A Tricentennial Tribute*. London: Imperial College Press, 2010.

Dehaene, Stanislas. "Precis of *The Number Sense*." *Mind and Language*, February 2001, 16-36.

Devlin, Keith. *The Language of Mathematics: Making the Invisible Visible*. New York: W.H. Freeman & Co., 1998.

Devlin, Keith. "The Most Beautiful Equation in Mathematics." *Wabash Magazine*, Winter/Spring 2002. http://www.wabash.edu/magazine/2002/WinterSpring2002/mostbeautiful.html.

Devlin, Keith. "Will Cantor's Paradise Ever Be of Practical Use?" Devlin's Angle. June 3, 2013. http://devlinsangle.blogspot.com/2013/06/will-cantors-paradise-ever-be-of.html.

Dudley, Underwood. "Is Mathematics Necessary?" *The College Mathematics Journal*, November 1997, 360-364. http://www.public.iastate.edu/~aleand/dudley.html.

Dunham, William. *Euler: The Master of Us All*. Washington, D.C.: The Mathematical Association of America, 1999.

Dunham, William. *Journey through Genius: The Great Theorems of Mathematics*. New York: Penguin Books USA, 1991.

Dunham, William, Editor. *The Genius of Euler: Reflections on His Life and Work*. Washington, D.C.: The Mathematical Association of America, 2007.

Euler, Leonhard. *Letters of Euler on Different Subjects in Physics and Philosophy Addressed to a German Princess*, trans. from the French by Henry Hunter. London: Murray and Highley, 1802.

Fellmann, Emil A. *Leonhard Euler*. Translated by Erika Gautschi and Walter Gautschi. Basel: Birkhäuser Verlag, 2007.

Fleron, Julian, with Volker Ecke, Philip K. Hotchkiss, and Christine von Renesse. *Discovering the Art of Mathematics: The Infinite*. Westfield, Mass.: Westfield State University, 2015. https://www.artofmathematics.org/books/the-infinite.

Gardner, Martin. "Is Mathematics for Real?" *The New York Review of Books*, August 13, 1981. http://www.nybooks.com/articles/1981/08/13/is-mathematics-for-real/.

Gardner, Martin. "Mathematics Realism and Its Discontents." *Los Angeles Times*, Oct. 12, 1997. http://articles.latimes.com/1997/oct/12/books/bk-44915.

Gardner, Martin. *The Magic and Mystery of Numbers*. New York: Scientific American, 2014.

Gray, Jeremy John. "Carl Friedrich Gauss." *Encyclopedia Britannica*, 2007. https://www.britannica.com/biography/Carl-Friedrich-Gauss.

Hardy, G. H. *A Mathematician's Apology*. Cambridge: Cambridge University Press, 1940.

Hatch, Robert A. "Sir Isaac Newton." *Encyclopedia Americana*, 1998. http://users.clas.ufl.edu/ufhatch/pages/01-courses/current-courses/08sr-newton.htm.

Hersh, Reuben. "Reply to Martin Gardner." *The Mathematical Intelligencer*, 23 (1), 2001, 3-5.

Hersh, Reuben. *What Is Mathematics, Really?* New York: Oxford University Press, 1997.

Horner, Francis. "Memoir of the Life and Character of Euler by the Late Francis Horner, Esq. M.P." In *Elements of Algebra* by Leonhard Euler, trans. from the French by Rev. John Hewlett. London: Longman, Orme, and Co., 1840.

Keynes, John Maynard. "Newton, the Man." Royal Society of London Lecture, 1944. http://www-history.mcs.st-and.ac.uk/Extras/Keynes_Newton.html.

Kline, Morris. *Mathematical Thought from Ancient to Modern Times*. New York: Oxford University Press, 1972.

Klyve, Dominic. "Darwin, Malthus, Süssmilch, and Euler: The Ultimate Origin of the Motivation for the Theory of Natural Selection." *Journal of the History of Biology*, Summer 2014. https://www.ncbi.nlm.nih.gov/pubmed/23948780.

Lakoff, George, and Rafael E. Núñez. *Where Mathematics Comes From: How the Embodied Mind Brings Mathematics into Being*. New York: Basic Books, 2000.

Lemonick, Michael D. "The Physicist as Magician." *Time*, Dec. 7, 1992.

Maor, Eli. *e: The Story of a Number*. Princeton: Princeton University Press, 1994.

Martin, Vaughn D. "Charles Steinmetz, The Father of Electrical Engineering." *Nuts and Volts*, April 2009. http://www.nutsvolts

.com/uploads/magazine_downloads/Steinmentz_Father_of_Elec_Engineering.pdf.

Martinez, Alberto A. *The Cult of Pythagoras: Math and Myths*. Pittsburgh: University of Pittsburgh Press, 2012.

Mazur, Barry. *Imagining Numbers: (Particularly the Square Root of Minus Fifteen)*. New York: Picador, 2003.

Moreno-Armella, Luis. "An Essential Tension in Mathematics Education." *ZDM Mathematics Education*, August 2014. http://link.springer.com/article/10.1007/s11858-014-0580-4.

Nahin, Paul J. *An Imaginary Tale: The Story of* $\sqrt{-1}$. Princeton: Princeton University Press, 1998.

Nahin, Paul J. *Dr. Euler's Fabulous Formula: Cures Many Mathematical Ills*. Princeton: Princeton University Press, 2006.

O'Connor John J., and Edmund F. Robertson. *MacTutor History of Mathematics Archive*. St. Andrews: University of St. Andrews, Scotland, 2016. http://www-history.mcs.st-and.ac.uk.

Ofek, Haim. *Second Nature: Economic Origins of Human Evolution*. Cambridge: Cambridge University Press, 2001.

Paulos, John Allen. "Review of *Where Mathematics Comes From*." *The American Scholar*, Winter 2002. https://math.temple.edu/~paulos/oldsite/lakoff.html.

Pólya, George. *Mathematics and Plausible Reasoning, Volume 1: Induction and Analogy in Mathematics*. Princeton: Princeton University Press, 1990.

Preston, Richard. "The Mountains of Pi." *The New Yorker*, March 2, 1992.

Reeder, Patrick J. Internal Set Theory and Euler's Introductio in Analysin Infinitorum, Ohio State University Master's Thesis, 2013. https://etd.ohiolink.edu/pg_10?0::NO:10:P10_ACCESSION_NUM:osu1366149288.

Reymeyer, Julie. "A Mathematical Tragedy." Science News, Feb. 25, 2008. https://www.sciencenews.org/article/mathematical-tragedy.

Roh, Kyeong Hah. "Students' Images and Their Understanding of Definitions of the Limit of a Sequence." *Educational Studies in Mathematics*, Vol. 69, 2008, 217-233.

Roy, Ranjan. "The Discovery of the Series Formula for π by Leibniz, Gregory, and Nilakantha." *Mathematics Magazine*, December 1990, 291-306.

Russell, Bertrand. *Mysticism and Logic: And Other Essays*. London: George Allen & Unwin Ltd., 1959.

Sandifer, C. Edward. "Euler Rows the Boat." *Euler at 300: An Appreciation*. Washington, D.C.: The Mathematical Association of America, 2007, 273-280.

Sandifer, C. Edward. "Euler's Solution of the Basel Problem—The Longer Story." *Euler at 300: An Appreciation*. Washington, D.C.: The Mathematical Association of America, 2007, 105-117.

Sandifer, C. Edward. "How Euler Did It: Venn Diagrams." The Euler Archive, January 2004. http://eulerarchive.maa.org/hedi/HEDI-2004-01.pdf.

Seife, Charles. *Zero: The Biography of a Dangerous Idea*. New York: Penguin Books, 2000.

Sinclair, Nathalie, David Pimm, and William Higginson, editors, *Mathematics and the Aesthetic: New Approaches to an Ancient Affinity*. New York: Springer Science+Business Media, 2006.

Steinmetz, Charles P. "Complex Quantities and Their Use in Electrical Engineering." *AIEE Proceedings of International Electrical Congress*, July 1893, 33–74.

Stewart, Ian. "Gauss." *Readings from Scientific American: Scientific Genius and Creativity*. New York: W.H. Freeman & Co., 1987.

Stewart, Ian. *Taming the Infinite: The Story of Mathematics*. London: Quercus Publishing, 2009.

Tall, David. *A Sensible Approach to the Calculus*. University of Warwick, 2010. homepages.warwick.ac.uk/staff/David.Tall/pdfs/dot2010a-sensible-calculus.pdf.

Toeplitz, Otto. *The Calculus: A Genetic Approach*. Chicago: University of Chicago Press, 1963.

Truesdell, Clifford, "Leonhard Euler, Supreme Geometer." *The Genius of Euler: Reflections on His Life and Works*. Washington, D.C.: The Mathematical Association of America, 2007, 13-41.

Tsang, Lap-Chuen. *The Sublime: Groundwork towards a Theory*. Rochester: University of Rochester Press, 1998.

Tuckey, Curtis, and Mark McKinzie. "Higher Trigonometry, Hyperreal Numbers, and Euler's Analysis of Infinities." *Mathematics Magazine*, December 2001. http://www.maa.org/sites/default/files/pdf/upload_library/22/Allendoerfer/2002/0025570x.di021222.02p0075s.pdf.

Wells, David. "Are These the Most Beautiful?" *The Mathematical Intelligencer*, 1990. https://www.gwern.net/docs/math/1990-wells.pdf.

Westfall, Richard S. "Sir Isaac Newton: English Physicist and Mathematician." In *Encyclopaedia Britannica* online. Chicago: Encyclopaedia Britannica Inc. http://www.britannica.com/biography/Isaac-Newton.

Wigner, Eugene. "The Unreasonable Effectiveness of Mathematics in the Natural Sciences." *Communications in Pure and Applied Mathematics*, February 1960, 1-14.

Zeki, Semir, John Paul Romaya, Dionigi M.T. Benincasa, and Michael F. Atiyah. "The Experience of Mathematical Beauty and Its Neural Correlates." *Frontiers in Human Neuroscience*, February 13, 2014. http://journal.frontiersin.org/article/10.3389/fnhum.2014.00068/full.

NOTES

INTRODUCTION

1 "So a few years ago..." The survey, which I first saw in 2014, was conducted in 1988 by David Wells and published in 1990 in *The Mathematical Intelligencer*.

3 "It's just that, inexplicably..." The calc books in question are Tom Apostol's *Calculus*, volumes I and II, published in 1969. They're still considered among the most rigorous, comprehensive calculus texts out there. Ironically, they're highly regarded partly because they play up math history. Volume I even includes a mini-hagiography of Euler noting that he "discovered one beautiful formula after another." Why don't they highlight what is widely regarded as the most beautiful one of all? I'll probably never know—Apostol, a former Caltech math professor, died in early 2016.

3 "Novelist Nicholson Baker memorably described..." Baker (2013).

4 "And I knocked it down." Bouwsma (1965), v.

5 "Unlike the physics or chemistry or engineering of today..." Nahin (2006), xx.

CHAPTER 1: GOD'S EQUATION

7 "Leonhard Euler seemed as curious..." Calinger (2015).

11 "Euler was the Enlightenment's greatest mathematician..." As Truesdell (2007) notes, Euler has long been known as one of

history's greatest mathematicians, but only in recent decades have historians recognized him as his century's leading physicist too.

11 "Science historian Clifford Truesdell has estimated..." Quoted in the Euler Archive, "Europe in the 18th Century." http://eulerarchive.maa.org.

12 "As mathematician William Dunham has noted..." Dunham (1999), 176.

12 "In 1752, for instance" Sandifer (2007).

12 "His ideas on how to construct achromatic lenses..." Calinger (2015), 383.

12 "He even proposed a design for a logic machine..." Sandifer (2004).

12 "...his writings clearly influenced Kant's metaphysics." Calinger (2015), 469.

12 "His work on the mathematics of population growth..." Klyve (2014).

12 "He had a hand in solving..." Calinger (2015), 389.

13 "Today, the mathematics he pioneered..." Stewart (2009), 104; and Nahin (2006).

13 "...perhaps only Voltaire..." Calinger (2015), 531.

14 "Mathematics textbooks call it Euler's formula." Euler didn't publish the equation in the form shown here, which apparently first appeared in the nineteenth century. But in 1728, he wrote an equivalent formula when demonstrating a way to calculate the areas of circles. In 1748, he proved a general equation from which his famous formula follows as a specific case in the *Introductio in analysin infinitorum* (Introduction to the Analysis of the Infinite)—this proof is detailed in Appendix 1. So Euler is recognized as the formula's primary author. However, some other mathematicians, including Johann Bernoulli, Euler's mentor, came close to discovering it. And in 1714, England's Roger Cotes developed an equation from which it could have been readily extracted. Cotes apparently didn't see the full implications of what he'd done, however, and he died two years later, at age 33.

14 "Benjamin Peirce, considered America's first world-class mathematician..." Quoted in Archibald (1925).
14 "Like a Shakespearean sonnet..." Devlin (2002).
14 "Richard Feynman was briefer..." Nahin (2006).
14 "And in 2014..." Zeki, Romaya, Benincasa, and Atiyah (2014).

CHAPTER 2: A CONSTANT THAT'S ALL ABOUT CHANGE

21 "...the symbolic comprehension of the infinite..." Quoted in Fleron (2015), 21.
21 "Rigorously defining *e* requires..." Euler and his contemporaries knew that affixing a numerical value to *e* involves a calculation that, in principle, is infinitely long. Specifically, it entails adding (or, alternatively, multiplying together) an infinite number of similar-looking fractions. Euler was a dazzling master at manipulating infinite sums and products to make important math advances. But in his day, the conceptual machinery wasn't available to do such manipulations without some fast and loose moves—ones that could lead to absurd results in certain situations. Even Euler, for all his genius, went astray at times because of this. As explained in Appendix 1, a carefully worded definition of limits, along with some other concepts that were developed after his era, helped eliminate the dubious maneuvering.
26 "...this is a high-altitude flyover..." Mathematician Arthur Benjamin offers a concise, readable introduction to calculus in *The Magic of Math: Solving for x and Figuring Out Why*.
28 "It required two great mathematicians..." While Newton and Leibniz are credited with inventing calculus, others came up with key pieces of the grand synthesis that became calculus, including Pierre de Fermat, Bonaventura Cavalieri, John Wallis, and James Gregory. Later mathematicians, prominently including Euler, helped flesh out and systematize the limited initial formulations of it by Newton and Leibniz. It seems that

the history of ideas is almost always more complicated than it appears at first glance.

29 "The latter has been empirically confirmed..." See, for example, Allain (2009).

31 "...followers of Pythagoras..." Martinez (2012).

CHAPTER 3: IT EVEN COMES DOWN THE CHIMNEY

33 "...it's an irrational number..." O'Connor and Robertson (2016), "Johann Heinrich Lambert."

33 "...by German mathematician Carl Louis Ferdinand von Lindemann..." Ibid., "Carl Louis Ferdinand von Lindemann."

33 "...it's a transcendental number..." Interestingly, Euler's formula played a key role in von Lindemann's proof that π is transcendental—the most beautiful equation's implications were deeper than anyone realized during Euler's time.

34 "...by French mathematician Charles Hermite." O'Connor and Robertson (2016), "Charles Hermite."

35 "...when French mathematician Joseph Liouville..." Ibid., "Joseph Liouville."

35 "Surely the population..." Wigner (1960).

35 "...this mysterious 3.14159..." Quoted in Gardner (2014).

36 "Historians believe the first discoverer..." Roy (1990).

37 "...after all, $2 - 3 + 4$ equals 3..." Mathematically knowledgeable readers will realize that I've pulled a fast one here by supposing that infinite sums behave like finite ones when it comes to regrouping terms. They often don't, of course. But that wasn't clear during Euler's era, and it makes sense to omit it in this historical example of confusion about the infinite.

38 "...never give out in our thought." Clegg (2003), 30.

39 "May we not call them the ghosts..." Ibid., 124.

40 "...unsettling issues about the infinite once again." Ibid.

40 "The most boggling implications of his theory..." In the last chapter of his book *Journey through Genius: The Great Theorems of Mathematics*, mathematician William Dunham of-

fers a nice explanation of how Cantor proved that the irrationals have a larger degree of infinity than the rationals—although it's a profound result, it's no harder to understand than a Sudoku puzzle.

40 "...declared that Cantor's theory..." Devlin (2013).
40 "...of Cantor's theory of the infinite." Ibid.
41 "This is why its symbol..." O'Connor and Robertson (2016), "John Wallis."
42 "...some 4,000 years ago." Blatner (1997).
43 "...a hypothetical polygon with 24,576 sides." O'Connor and Robertson (2016), "Zu Chongzhi."
44 "...the problem was first posed in 1644..." Sandifer (2007), "Euler's Solution of the Basel Problem—The Longer Story."
45 "...amateur British mathematician William Shanks..." O'Connor and Robertson (2016), "William Shanks."
45 "One of the more notable records..." Preston (1992).

CHAPTER 4: THE NUMBER BETWEEN BEING AND NOT-BEING

49 "...claimed Italian mathematician Gerolamo Cardano." O'Connor and Robertson (2016), "Girolamo Cardano."
52 "In the 1500s..." Nahin (1998).
53 "...actually a camouflaged real number." Ibid.
54 "...between being and non-being." O'Connor and Robertson (2016), "Quotations by Gottfried Leibniz."
54 "Essentially throwing up his hands..." Quoted in Dunham (1999), 87.

CHAPTER 5: PORTRAIT OF THE MASTER

55 "Let's briefly veer off..." Unless otherwise noted, biographical details in this chapter are taken from Calinger (2015) and Fellmann (2007).
55 "My favorite description of Euler..." Thiebault's appealing description of Euler may not be accurate, according to math

historian Ronald Calinger, since the French linguist didn't know Euler when his children were young. And it was likely that Euler's wife mainly cared for the couple's children while her husband worked. Still, Thiebault's depiction accords with other, better established accounts of Euler's personality and domestic life.

56 "...as mathematician William Dunham has noted..." Dunham (1999), xx.

57 "...marveled twentieth-century French mathematician André Weil." Quoted in Dunham (1999), xv.

58 "...according to historian Eric Temple Bell." Bell (1986).

58 "Bell added, because he possessed..." Ibid., 140.

58 "Still, when recently going over Euler's derivations..." William Dunham presents lucid expositions of a selection of Euler's important theorems in *Euler: The Master of Us All*. A person who's completed a basic calculus course could follow most of it if willing to break into a minor sweat now and then.

58 "An ordinary genius is a fellow..." Quoted in Lemonick (1992).

61 "...as historian Clifford Truesdell sardonically put it." Truesdell (2007), 22.

63 "If the stories told about Archimedes are true..." O'Connor and Robertson (2016), "Archimedes of Syracuse."

64 "William Whiston, who assisted Newton..." Quoted in Keynes (1944).

64 "As historian Robert A. Hatch put it..." Hatch (1998).

65 "To top it off..." Ibid.

65 "After they quarreled..." Cajori (1899).

65 "One such case concerned..." O'Connor and Robertson (2016), "János Bolyai."

67 "Gauss never met her..." Quoted in Rehmeyer (2008).

67 "...because he'd feared getting attacked..." Burris (2009), "Letter from Gauss to Bessel, January 27, 1829."

67 "...remarked British mathematician Ian Stewart." Stewart (1987).

67 "Gauss's contemporaries less diplomatically..." Gray (2007).

CHAPTER 6: THROUGH THE WORMHOLE

73 "When author Alex Bellos surveyed..." Bellos (2014), 19.

75 "In other words, when the three enigmatic numbers..." After I wrote a draft of this book, I ran across a 2014 article in *Wired* by another fan of Euler's identity, Lee Simmons, who also used a wormhole metaphor to describe Euler's formula. So as not to seem derivative, I tried to think of an alternative metaphor. But then I decided to stick with wormhole and to note that this is an example of convergent evolution in memespace, analogous to the evolving of sonar-like echolocation in bats, whales, certain birds, and shrews.

CHAPTER 7: FROM TRIANGLES TO SEESAWS

77 "Though wonderfully ingenious..." Arthur Benjamin offers a straightforward, calculus-based derivation of Euler's identity in *The Magic of Math*. (By the way, don't miss Benjamin's "mathemagics" TED talk on YouTube, a crowd-pleaser that's been viewed over seven million times.)

77 "...a no-choke mini-primer..." If you want a more comprehensive tutorial on trig, the online Khan Academy (www.khanacademy.org) is a good place to look.

77 "As he put it, functions are..." O'Connor and Robertson (2016), "History Topic: The Function Concept."

83 "a forerunner to defining trig functions..." Ibid., "Hipparchus of Rhodes."

CHAPTER 9: PUTTING IT TOGETHER

104 "De Moivre is credited with..." Ibid., "Abraham de Moivre."

105 "One of Euler's intriguing findings..." Euler's derivations of the infinite sums for cos θ and sin θ lie outside the conceptual

boundaries of this book. If you'd like to see them, William Dunham's book, *Euler: The Master of Us All*, covers them on pages 92–3.

105 "That is, he showed that..." Trig-related formulas involving infinite sums were known before Euler's time. Impressively, for instance, Indian mathematicians of the fifteenth century worked out a number of them.

109 "Here's the infinite sum..." Euler devised an elegant proof that the infinite sum $1 + x + x^2/2! + \cdots$ equals e^x. But this math fact was prefigured in the work of Isaac Newton.

CHAPTER 10: A NEW SPIN ON EULER'S FORMULA

114 "An acquaintance described him..." Branner and Johansen (1999).

115 "...a fellow surveyor once wrote of him." Ibid.

115 "Finally, in 1895..." A few mathematicians, in particular England's John Wallis, adumbrated Wessel's geometric interpretation by suggesting how to picture complex numbers geometrically on a two-dimensional plane. But these early beginnings didn't lead anywhere.

115 "A few years after Wessel's big idea..." O'Connor and Robertson (2016), "Jean Robert Argand."

123 "...devised this rule in the 1500s..." Ibid., "Rafael Bombelli."

137 "Math historians Edward Kasner and..." Quoted in Maor (1994), 153.

CHAPTER 11: THE MEANING OF IT ALL

139 "In 1892 he joined..." Steinmetz (1893).

139 "Delighted by animals that..." Martin (2009).

140 "In later life he was dubbed..." Nahin (1998), 137.

140 "Today, Euler's formula is a tool..." Nahin (2006) contains instructive examples of how Euler's formula is used in electrical engineering.

142 "a much-cited observation by the great British philosopher and mathematician Bertrand Russell..." Russell (1959).
144 "English mathematician G. H. Hardy..." Hardy (1940).
144 "...there is no permanent place in the world for ugly mathematics..." Hardy (1940), 84.
145 "...he wrote that the sublime..." Tsang (1998), 3.
147 "French chemical engineer, writer, and amateur mathematician..." Quoted in Wells (1990).
149 "Keith Devlin, the mathematician who compared..." Devlin (1998), 134.
150 "This view has been held..." Hardy (1940), 124.
150 "And on other days..." Mazur (2003), 70.
151 "He argued that 'if all...'" Gardner (1981).
152 "It takes enormous hubris..." Gardner (1997).
154 "Physicist Eugene Wigner famously dubbed..." Wigner (1960).
155 "'...which can thus be said to inhere in the universe,' he wrote." Paulos (2002).
155 "Based on this logic..." Ofek (2001).
156 "...to represent basic arithmetic knowledge." Dehaene (2001).
156 "He and colleague Marie Amalric..." Amalric and Dehaene (2016).
156 "Dehaene observes, for instance..." Dehaene (2001).
158 "Ramanujan was referring to the fact..." Bradley (2006).
162 "A memorable observation by German physicist..." O'Connor and Robertson (2016), "A Quotation by Heinrich Hertz."

APPENDIX 1: EULER'S ORIGINAL DERIVATION

165 "Euler presented his initial proof..." Written in Latin, as was customary at the time, the *Introductio*'s two volumes were first published in English in 1988 and 1990, translated by John D. Blanton. Here, I've relied on William Dunham's explication of Euler's derivation in his book, *Euler: The Master of Us All*, and on an English translation of the *Introductio* by Ian Bruce, a retired Australian physics professor who has

impressively translated several of Euler's major works and posted them at www.17centurymaths.com.

166 "In a 1950 lecture, historian Carl Boyer..." Boyer (1950).

167 "One reason is that revisionist mathematicians..." Examples include the papers by Mark McKenzie and Curtis Tuckey; Jacques Bair, et al.; and Patrick Reeder that are cited in the bibliography.

167 "Further, some math educators have argued..." See, for example, Luis Moreno-Armella's "An Essential Tension in Mathematics Education," listed in the bibliography. David Tall makes similar points in his book, *A Sensible Approach to Calculus*, also cited in the bibliography.

168 "Largely because of that..." Pólya (1990).

168 "Mathematician William Dunham has similarly noted..." Dunham (1999).

170 "...according to eminent math historian Morris Kline" Kline (1972), 619.

INDEX

ALICIA RUSSELL

David Stipp is an award-winning science writer whose work has appeared in *Scientific American*, *New York Times*, *Wall Street Journal*, *Science*, and other publications. The author of *The Youth Pill*, he lives in Arlington, Massachusetts.